Geodæsia

You are holding a reproduction of an original work that is in the public domain in the United States of America, and possibly other countries. You may freely copy and distribute this work as no entity (individual or corporate) has a copyright on the body of the work. This book may contain prior copyright references, and library stamps (as most of these works were scanned from library copies). These have been scanned and retained as part of the historical artifact.

This book may have occasional imperfections such as missing or blurred pages, poor pictures, errant marks, etc. that were either part of the original artifact, or were introduced by the scanning process. We believe this work is culturally important, and despite the imperfections, have elected to bring it back into print as part of our continuing commitment to the preservation of printed works worldwide. We appreciate your understanding of the imperfections in the preservation process, and hope you enjoy this valuable book.

TA
544
.L897
1720

GEODÆSIA;

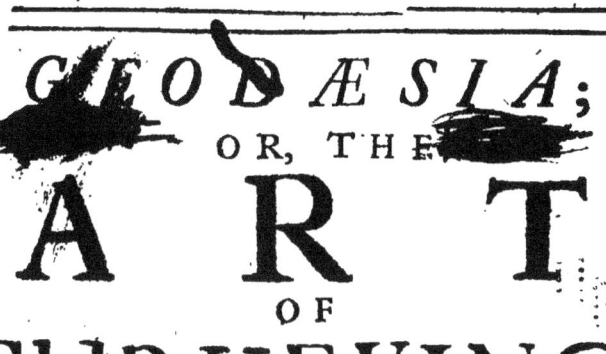

OR, THE ART OF SURVEYING

AND Measuring of LAND Made EASI.

SHEWING,

By Plain and Practical Rules, how to Survey, Protract, Cast up, Reduce or Divide any Piece of Land whatsoever; with New Tables for the ease of the Surveyor in Reducing the Measures of Land.

MOREOVER,

A more facile and sure Way of Surveying by the Chain, than has hitherto been Taught.

AS ALSO,

How to lay out New Lands in **America** or elsewhere: And how to make a Perfect Map of a River's Mouth or Harbour; with several other Things never yet Publish'd in our Language.

By *JOHN LOVE*.

The Third Edition with Additions.

LONDON:

Printed for *W. TAYLOR*, at the *Ship* and *Black Swan* in Pater-Noster-row, 1720.

Soth,
8435
math.
4-27-1923

THE PREFACE TO THE READER.

WHAT would be more ridiculous, than for me to go about to Praise an Art that all Mankind know they cannot live Peaceably without? It is near hand as ancient (no doubt on't) as the World: For how could Men set down to Plant, without knowing some Distinction and Bounds of their Land? But (Necessity being the Mother of Invention) we find the *Egyptians*, by reason of the *Nyles* over-flowing, which either washt away all their Bound Marks, or cover'd them over with Mud, brought this Measuring of Land first into an Art, and Honoured much the Professors of it. The great Usefulness, as well as the pleasant and delightful

The PREFACE.

lightful Study, and wholsom Exercise of which, tempted so many to apply themselves hereto, that at length in *Egypt* (as in *Bermudas*) every Rustick could Measure his own Land.

From *Egypt*, this Art was brought into *Greece*, by *Thales*, and was for a long time called *Geometry*; but that being too comprehensive a Name for the Mensuration of a Superficies only, it was afterwards called *Geodæsia*; and what Honour it still had continued to have among the Antients, needs no better Proof than *Plato*'s ἀγεωμέτρητος ὐδεὶς εἰσίτω And not only *Plato*, but most, if not all the Learned men of these times, refused to admit any into their Schools, that had not been first entred in the *Mathematicks*, especially *Geometry* and *Arithmetick*. And we may see, the great Monuments of Learning built on these Foundations, continuing unshaken to this Day, sufficiently demonstrate the Wisdom of the Designers, in chusing *Geometry* for their Ground-Plot.

Since which, the *Romans* have had such an Opinion of this sort of Learning, that they concluded that Man to be incapable

of

The PREFACE.

of Commanding a Legion, that had not at least so much *Geometry* in him, as, to know how to Measure a Field. Nor did they indeed either respect Priest or Physitian, that had not some Insight in the *Mathematicks*.

Nor can we complain of any failure of Respect given to this Excellent Science, by our Modern Worthies, many Noblemen, Clergymen, and Gentlemen affecting the Study thereof: So that we may safely say, none but Unadvised Men ever did, or do now speak evil of it.

Besides the many Profits this Art brings to Man, it is a Study so pleasant, and affords such Wholesome and Innocent Exercise, that we seldom find a Man that has once entred himself into the Study of *Geometry* or *Geodæsia*, can ever after wholly lay it aside; so natural is it to the Minds of Men, so pleasingly insinuating, that the *Pythagoreans* thought the Mathematicks to be only a Reminiscience, or calling again to Mind things formerly learned.

But no longer to light Candles to see the Sun by, let me come to my Business, which

The PREFACE.

is to speak something concerning the following Book; and if you ask, why I write a Book of this nature, since we have so many very good ones already in our own Language? I answer, because I cannot find in those Books, many things, of great consequence, to be understood by the Surveyor. I have seen Young men, in *America*, often non-plus'd so, that their Books would not help them forward, particularly in *Carolina*, about Laying out Lands, when a certain quantity of Acres has been given to be laid out five or six times as broad as long. This I know is to be laught at by a Mathematician; yet to such as have no more of this Learning, than to know how to Measure a Field, it seems a difficult Question: And to what Book already Printed of Surveying shall they repair to, to be resolved?

Also concerning the *Extraction of the Square Root*; I wonder that it has been so much neglected by the Teachers of this Art, it being a Rule of such absolute necessity for the Surveyor to be acquainted with. I have taught it here as plainly as I could devise, and that according to the Old way,

The PREFACE.

way, verily believing it to be the Best, using fewer Figures, and once well learned, charging less the memory than the other way.

Moreover, the Sounding the Entrance of a River, or Harbour, is a Matter of great Import, not only to Seamen, but to all such as Seamen live by; I have therefore done my endeavour to teach the Young Artist how to do it, and draw a fair Draught thereof.

Many more things have I added, such as I thought to be New, and Wanting; for which I refer you to the Book it self.

As for the Method, I have chose that which I thought to be the easiest for a Learner; advising him first to learn some Arithmetick, and after, teaching him how to Extract the Square-Root. But I would not have any Neophyte discouraged, if he find the *First* Chapter too hard for him; for let him rather skip it, and go to the *Second* and *Third* Chapters, which he will find so easie and delightful, that I am perswaded he will be encouraged to conquer the Difficulty of learning that one Rule in the *First* Chapter.

The PREFACE.

From *Arithmetick*, I have proceeded on to teach so much *Geometry* as the Art of *Surveying* requires. In the next place I have shewed by what Measures Land is Surveyed, and made several Tables for the Reducing one sort of Measure into another.

From which I come to the Description of Instruments, and how to Use them; wherein I have chiefly insisted on the Semicircle, it being the best that I know of.

The *Sixth* Chapter teacheth how to apply all the foregoing Matters together, in the Practical Surveying of any Field, Wood, &c. divers Ways, by divers Instruments; and how to lay down the same upon Paper. Also at the end of this Chapter I have largely insisted on, and by new and easie ways, taught Surveying by the Chain only.

The *Seventh*, *Eighth*, *Ninth*, *Tenth* and *Eleventh* Chapter, teach how to cast up the Contents of any Plot of Land; How to lay out New Lands; How to Survey a Mannor, County or Country: Also, how to Reduce, and Divide Lands, *Cum multis aliis*.

The PREFACE.

The *Twelfth* Chapter consists wholly of *Trigonometry*.

The *Thirteenth* Chapter is of Heights and Distances, including amongst other things, how to make a Map of a River or Harbour. Also how to convey Water from a Spring-head, to any appointed Place, or the like.

Lastly, At the end of the Book, I have a Table of Northing or Southing, Easting or Westing; or (if you please to call it so) A Table of Difference of Latitude and Departure from the Meridian, with Directions for the Use thereof. Also a Table of Sines and Tangents, and a Table of Logarithms.

I have taken Example from Mr. *Holwell* to make the Table of *Sines* and *Tangents*, but to every Fifth Minute, that being nigh enough in all Sense and Reason for the Surveyor's Use; for there is no Man, with the best Instrument that was ever yet made, can take an Angle in the Field nigher, if so nigh, as to Five Minutes.

All

The PREFACE.

All which I commend to the Ingenious Reader, wishing he may find Benefit thereby, and desiring his favourable Reception thereof accordingly, I conclude,

READER,

Your Humble Servant,

J. L.

ADVERTISEMENT.

Such Persons as have occasion for the Instruments mentioned in this Book, or any other Mathematical Instruments whatsoever, may be furnished with the same at Reasonable Rates, by *John Rowley*, Instrument-Maker, at his Shop under the Dial of St. *Dunstan*'s Church in *Fleet-street*, *London*.

THE

THE CONTENTS.

CHAP. I.

	Page
OF Arithmetick in general	1
How to Extract the { Vulgar Arithmetick	2
Square Root, by { The Logarithms	7

CHAP. II.

Geometrical Definitions.

	Page
Shewing what is meant by { A Point	9
A Line	ibid.
An Angle	ibid.
A Perpendicular	10
A Triangle	11
A Square	12
A Parallelogram	ibid.
A Rhombus and Rhomboides	ibid.
A Trapezia	ibid.
An Irregular Figure	13
A Regular Polygon, as Pentagon, Hexagon, &c.	14
A Circle, with what thereto belongs	ibid.
A Superficies	15
Parallel-Lines	16
Diagonal-Lines	ibid.

CHAP.

The CONTENTS.

CHAP. III.

Geometrical Problems.

                              Page

1. HOW to make a Line Perpendicular to another two ways   17
2. How to raise a Perpendicular upon the end of a Line two ways   18
3. How from a Point assigned, to let fall a Perpendicular upon a Line given   20
4. How to Divide a Line into any Number of Equal Parts   21
5. How to make an Angle equal to any other Angle given   22
6. How to make Lines Parallel to each other   23
7. How to make a Line Parallel to another Line, which must also pass through a Point assigned   24
8. Three Lines being given, how to make thereof a Triangle   ibid.
9. How to make a Triangle equal to a Triangle given   25
10. How to make a Square Figure   26
11. How to make a Long Square or Parallelogram ibid.
12. How to make a Rhomubs or Rhomboides   27
13. How to make Regular Polygons, as Pentagons, Hexagons, Heptagons, &c.   28
14. Three Points being given, how to make a Circle, whose circumference shall pass through the three Points   32
15. How to make an Ellipsis, or Oval, several ways 33
16. How to Divide a given Line into two Parts, which shall be in such Proportion to each other as two given Lines   36

17. *Three*

The CONTENTS.

17. *Three Lines being given; to find a Fourth in Proportion to them.* 37

CHAP. IV.

Of Measures in general.

1. Of Long Measure, shewing by what kind of Measures Land is Surveyed; and also how to Reduce one sort of Long Measure into another 39
A General Table of Long Measure ibid.
A Table shewing how many Feet and Parts of a Foot; also how many Perches and Parts of a Perch are contained in any number of Chains and Links from one Link to an hundred Chains 41
A Table shewing how many Chains, Links and Parts of a Link; also how many Perches and Parts of a Perch, are contained in any number of Feet, from 1 Foot to 10000 44
II. Of Square Measure, shewing what it is; and how to Reduce one sort into another 46
A General Table of Square Measure 47
A Table, shewing the Length and Breadth of an Acre, in Perches, Feet, and Parts of a Foot 49
A Table to turn Perches into Acres, Roods and Perches 53

CHAP.

The CONTENTS.

CHAP. V.

Of Instruments and their Use.

                                                              Page

OF the Chain   54
Of Instruments for the taking of an Angle in the Field   56
To take the quantity of an Angle in the Field by
        Plain Table   57
        Semi-circle   58
        Circumferentor, &c. several ways ibid.
Of the Field-Book   61
Of the Scale, with several Uses thereof; and how to make a Line of Chords   62, &c.
Of the Protractor   68

CHAP. VI.

HOW to take the Plot of a Field, at one Station, in any place thereof; from whence you may see all the Angles by the Semi-circle; and to Protract the same 71
How to take the Plot of the same Field, at one Station, by the Plain Table   74
How to take the Plot of the same Field, at one Station, by the Semicircle, either with the help of the Needle and Limb both together, or by the help of the Needle only   ibid.
How, by the Semi-circle, to take the Plot of a Field, at one Station, in any Angle thereof, from whence the other Angles may be seen, and to Protract the same   76

How

The CONTENTS.

               Page

How to take the Plot of a Field, at two Stations, provided from either Station you may see every Angle, and measuring only the Stationary Distance, also to Protract the same   79, 82, &c.

How to take the Plot of a Field, at two Stations, when the Field is so Irregular, that from one Station you cannot see all the Angles   83

How to take the Plot of a Field, at one Station, in an Angle (so that from that Angle you may see all the other Angles) by measuring round about the said Field   86

How to take the Plot of the foregoing Field, by measuring one Line only; and taking Observations at every Angle   88

How to take the Plot of a large Field or Wood, by measuring round the same; and taking Observations at every Angle by the Semi-circle   90

When you have Surveyed after this manner, how to know, before you go out of the Field, whether you have wrought true or not   94

Directions how to Measure Parallel to a Hedge, when you cannot go in the Hedge it self: And also in such case, how to take your Angles   95

How to take the Plot of a Field or Wood, by observing near every Angle, and Measuring the Distance between the Marks of Observation, by taking in every Line two Off-sets to the Hedge   97

An easier way to do the same, by taking only one Square and many Off-sets

How by the help of the Needle to take the Plot of a large Wood, by going round the same, and making use of that division of the Card that is numbred with four 90s.

The CONTENTS.

Page

or Quadrants; and two ways how to Protract the same, and examine the Work - 103, &c.

How by the Chain only, to take an Angle in the Field 111

How by the Chain only, to survey by a Field by going round the same 113

The Common way taught by Surveyors, for taking the Plot of the foregoing Field 116

How to take the Plot of a Field, at one Station, in any part thereof, from whence all the Angles may be seen, by the Chain only 119

CHAP. VII.

How to cast up the Contents of a Plot of Land.

OF the Square and Parallelogram 120
Of Triangles 123
To find the Content of a Trapezia 125
How to find the Content of an Irregular Plot, consisting of many Sides and Angles 127
How to find the Content of a Circle, or any Portion thereof 128
How to find the Content of an Oval 130
How to find the Content of Regular Polygons, &c. 131

CHAP.

The CONTENTS.

CHAP. VIII.

Of Laying out New Lands.

A Certain quantity of Acres being given, how to lay out the same in a Square Figure   132

How to lay out any given quantity of Acres in a Parallelogram, whereof one side is given   133

How to lay out a Parallelogram that shall be four, five, six, or seven times, &c. longer than broad   ibid.

How to make a Triangle that shall contain any number of Acres, being confined to a given Base   134

How to find the Length of the Diameter of a Circle, that shall contain any number of Acres required   136

CHAP. IX.

Of Reduction.

HOW to Reduce a large Plot of Land, or Map, into a lesser compass, according to any given Proportion. Or contrary, how to enlarge one, three several ways.   137

How to change Customary Measure into Statute; and contrary   141

Knowing the Content of a piece of Land, to find out what Scale it was Plotted by.   ibid.

(a)   CHAP

The CONTENTS.

CHAP. X.

Inſtructions for Surveying A Mannor, County, or Country. 142

CHAP. XI.

Of Dividing Lands.

How to Divide a Triangular Piece of Land ſeveral ways 146
How to reduce a Trapezia into a Triangle, by Lines drawn from any Angle thereof. Alſo how to reduce a Trapezia into a Triangle, by Lines drawn from a Point aſſigned in any Side thereof. 149
How to Reduce a Five-ſided Figure into a Triangle, and to Divide the ſame 151
How to Divide an Irregular Plot of any number of Sides, according to any given Proportion, by a ſtreight Line throught it. 153
An eaſier Way to do the ſame; with two Examples 155
How to Divide a Circle, according to any Proportion, by a Circle Concentrick with it. 158

CHAP.

The CONTENTS.

CHAP. XII.

Trigonometry, 159, &c.

THis Chapter shews first the Use of the Tables of Sines and Tangents. And Secondly, contains Ten Cases for the Mensuration of Right-lin'd Triangles, very necessary to be understood by the Surveyor.

CHAP. XIII.

Of Heights and Distances.

HOW to take the Height of a Tower, Steeple, Tree, or any such thing 180
How to take the Height of a Tower, &c when you cannot come nigh the foot thereof 183
How to take the Height of a Tower, &c. when the Ground either riseth or falls. 185
How to take the Horizontal Line of a Hill 189
How to take the Rocks or Sands at the Entrance of a River or Harbour, and to Plot the same 191
How to know whether Water may be made to run a from Spring-Head to any appointed Place 194
A Table of Northing, or Southing, Easting or Westing.
A Table of Logarithms.
A Table of Artificial Sines and Tangents

BOOKS Printed for WILLIAM TAYLOR, at the SHIP and BLACK-SWAN, in PATER-NOSTER-ROW.

1. *Trigonometry* Improv'd, and *Projection* of the *Sphere* made Easie, teaching the Projection of the Sphere *Orthographick* and *Stereographick*, as also *Trigonometry* Plain and *Spherical*, with plain and intelligible Reasons for the various and most useful Methods, both in *Projection* and *Calculation*; with the Application of the whole to *Astronomy*, *Dialing* and *Geography*.

2. The *London* Accomptant, or Arithmetick in all its Parts, *viz.* both in whole Numbers, and Fractions, Vulgar and Decimal. with the Extraction of the Square and Cube Root, with the Reasons of the Operations, demonstrated from Copper-Plates. These two by *Henry Wilson*.

3. The Elements of *Euclid*, with select Theorems out of Archimedes, by that learned *Jesuit Andrew Tacquet*, to which are added Practical Corollaries, shewing the use of many of the Propositions, by *William Whiston*, 8vo. Price 4s. 6d.

4. *Astronomical Lectures*, read in the publick Schools at *Cambridg*, by Mr. *Whiston*, 8vo. Price 6s.

5. Sir *Isaac Newton*'s Mathematical Philosophy, more easily Demonstrated. and Dr. *Halley*'s Account of Comets, Illustrated in 40 Lectures, read at *Cambridge*, 8vo Price 6s.

6. *Astronomical* Principles of Religion, Natural, and Reveal'd, in 9 parts, 8vo. Price 5s.

7. *Clavis Usuræ*, or a Key to Interest, both Simple and Compound, with rules for Estimating the Value of Annuities, or Leases, and Insurance for Lives, & . also Rules for Rebate, Discompt, Equation of Payments &. by *J. Ward* Price 2s.

8. *Universal Arithmetik*, or, a Treatise of Arithmetical Composition and Resolution, to which is added, Dr. *Halley*'s Method of finding the Root of Equations Arithmetically ; Translated by the late Mr. *Ralphson*, and Corrected and Revised by Mr. *Cunn*.

9. Mr. *Wingate*'s Arithmetick, Containing a plain and Familiar Method, for Attaining the Knowledge, and Practice of Arithmetick,

These and Most other *Mathematical* Books, are to had of *William Taylor*, at the *Ship* and *Black Swan*, in *Pater Noster Row*.

GEODÆSIA:

OR, THE
ART
OF
Measuring Land, &c.

CHAP. I.
Of ARITHMETICK.

IT is very necessary for him that intends to be an Artist in the Measuring of Land, to begin with Arithmetick, as the Ground-work and Foundation of all Arts and Sciences Mathematical: and at least not to be ignorant of the five first and principal Rules thereof, *viz. Numeration, Addition, Subtraction, Multiplication* and *Division*: Which supposing every Person, that applies himself to the Study of this Art to be skilled in; or if not, referring him to Books or Masters, (every

B   where

Of Arithmetick.

where to be found) to learn: I shall name a sixth Rule, as necessary, (if not more) to be understood by the Learner; which is the Extraction of the Square Root; without which (though seldom mentioned by Surveyors in their Writings) a Man can never attain to a competent Knowledge in the Art: I shall not therefore think it unworthy my Pains (though perhaps other Men have better done it before me) to shew you easily and briefly how to do it.

How to Extract the Square Root.

In the first place it is convenient to tell you what the Square Root is: It is to find out of any Number propounded a lesser Number, which lesser Number being multiplyed in it self, may produce the Number propounded. As for Example, suppose 81 be a Number given me, I say 9 is the Root of it, because 9 multiplyed in it self, *viz.* 9 times 9, is 81. Now 8 could not be the root, for 8 times 8 is but 64: nor could 10, for 10 times 10 is 100, therefore I say 9 must needs be the Root, because multiplyed in it self, it makes neither more nor less, but just the Number propounded, *viz.* 81.

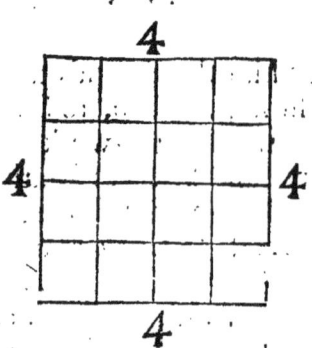

Again, suppose 16 be the number given, I say the Root of it is 4, because 4 multiplyed in it self makes 16. For your better understanding see this Figure, which is a great Square, containing 16 little Squares; any side of which great Square contains 4 little Squares: which is called the Square Root. Or

Of Arithmetick.

Or, suppose a plain Square Figure be given you as this in the Margent, and it be required of you to divide it into 9 small Squares: Your Business is to know into how many Parts to divide any one of the Side Lines, which here must be into 3, and that is the Root required. But now how to do this readily is the thing I am now going to teach you. The Roots of all Square Numbers under 100, you have in your Multiplication Table, however since it is good for you to keep them in your Mind, take this small Table of them.

Roots	1	2	3	4	5	6	7	8	9
Squares	1	4	9	16	25	36	49	64	81

Here you see the Root of 25 is 5, the Root of 64 is 8, and so of the rest.

So far as 100 in whole Numbers, your Memory will serve you to find the Root; but if the Number propounded, whose Root you are to search out, exceed 100, than put a Point over the first Figure on the Right-hand, which is the place of Unites, and so proceeding to the Left-hand, miss the second Figure, and put a Point over the third, then missing the fourth, Point the fifth; and so (if there be never so many Figures in the Number) proceed on to the end, pointing every other Figure, as you may see here, and so many Points as there are, of so many Figures your Root will 1234567 consist, which is very material to remember: Then begin at the first Figure on the Left-hand that has a point over it, which will always be the first or second Figure, and search out the Root

B 2

of that one Figure, or both joyned together if there be two, and when you have found it, or the nigheſt leſs to it, which you may eaſily do by the Table above, or your own Memory, draw a little crooked Line, as in Diviſion, and there ſet it down. For Example, Let 144 be the Number whoſe Root I am to find; I ſet it down, and prick the Figures thus:

144(12
22

Then going to the firſt Figure on the Left-hand, that has a Prick over it, which is 1, and ſee what the Root of it is, which is 1 alſo; I therefore draw a crooked Line, as in the Margent, and ſet down 1 in the Quotient, then if 1 admitted of any Multiplication, I ſhould multiply it by itſelf, but ſince once 1 is but 1, I ſubtract it out of the firſt prick'd Figure on the Left-hand, and there remains 0, ſo that I cancel that firſt Figure, as having wholly done with it: If any thing had remained after the Subtraction, I ſhould have put the remainder over it. The next thing to be done, is to double what is already in the Quotient, which makes 2, which 2 I write down under the next Figure, viz. 4, which has no Point over it, and then ſee how oft I can have 2 in 4: Anſwer, twice; I therefore ſet down 2 in the Quotient, and 2 likewiſe under the next pointed Figure, which in this Example is 4, then that 22 which ſtands under the 44 muſt be multiplyed by the 2 in the Quotient, whoſe Product is 44, which Subtracted out of 44, there remains 0: But you may multiply and ſubtract together thus, twice 2 is 4, which I take out of 4, and there remains 0, then I cancel the firſt 4 and 2 to the Left-hand, as having done with them; then again, twice 2 is 4, which taken out of 4 leaves 0, and then I cancel the laſt 4 and 2, and the Queſtion

Of Arithmetick

stion is answered, for there is 12 in the Quotient, which is the Root of 144, which may easily be proved by multiplying 12 by 12.

Take another Example: Let the sum be 54756. First see what the Root of 5 is, which is 2, and place it in the Quotient, and under the first pointed Figure both, as you see here, then say two times 2 is 4, which taken out of 5, there remains one, and so have you done with the first Point. Next double the Quotient, which makes 4, and place it as you see here, under the Figure void of a Point, then see how many times 4 you can have in 14, answer 3 times, which 3 place both in the Quotient, and under the next pointed Figure, which is 7; then multiply and subtract, saying three times 4 is 12, which taken out of 14 leaves 2, which 2 write over the 4, and cancel both the 4 and the 1, as you do in Division: And three times 3 is 9, which taken out of 27, rests 18; which write over head, and cancel what Figures you have done with, no otherwise than in Division, and so have you done with the first two Points. Now for the third pointed Figure, or if there were never so many more of them, they are done altogether as the second: *viz.* Double again your Quotient, it makes 46, which put down as you see in the Margin, always observing this Rule, That the last Figure of the doubled Quotient, I mean that in the place of Unites, stand under the next, void of Points: And those of your Left-hand of him, *viz.* in the place of Tens or Hundreds, in order before him, as you do in Division, as you may see here: Then proceed, and say,

B 3   how

Of Arithmetick.

```
   1        how many times 46 can I have in 185,
  2·8       or rather how many times 4 in 18: here
54756 (23   Essay, as you do in Division, and see if
 3436      you can have it four times, remembring
   4       the 4 that must be put down under the
```
pointed Figure, and when you find you can have it four times, write it down in the Quotient, and also under your last pointed Figure; then say four times 4 is 16, out of 18, there rests 2, which
```
  12        write down, and cancel the 18 and 4.
 2·8·16     Again, four times 6 is 24, out of 25,
54756 (234  rests 1, which put down, and cancel the
 24364      2, 5, and 6. Again, four times 4 is 16,
   4        out of 16, rests 0: and so have you done,
```
and find the Root to be 234.

I'll add but one Example more for your practice. Let the Number, whose Root is required be 12345678, see the working of it.

```
              But in this you see there is a
    2·0        Fraction remains, and so there
   3·845·9    will be in most Numbers, for we
12345678 (3513 seldom happen upon a Number
   365·07·23   exactly Square: the Fractional
    77·2       Part must therefore thus be taken:
```
before you begin to extract, add to your Number given two Cyphers, if you desire to know but to the tenth part of an Unite; but if to an hundredth part add four Cyphers, if to a thousandth part of an Unite, add six Cyphers, and then work, as before, as if it was all one entire Number, and look how many Points were placed over the Number first given, so many places of Integers will be in the Root; the rest of the Root towards the Right-hand, will be the Numerator of a Decimal Fraction. For Example,

let

Of Arithmetick. 7

let 143 by the Number given to be extracted, and to know the Decimal Fraction as near as to the hundredth part of an Unite; I write it down as before, annexing four Cyphers to the end of it, as you see hereunder; and after having wrought it, there comes out in the Quotient 11.95, but because I had but two Points over the first Number given, *viz.* 123. I therefore at the end of two Figures in the Quotient put a Point, which parts the whole Number from the Fraction; that 11 on the Left-hand being Integers, and the 95 on the Right Centesms of an Unite, which you may either write as above, or thus, 11.$\frac{95}{100}$ if you please.

There are other ways taught by Arithmeticians for finding out the Square Root of any Number; but I know no way so concise as this, and after a little practice, so easy and ready, or to be wrought with as few Figures. To do it indeed by the Logarithms or Artificial Numbers, is very easy and pleasant, but Surveyers have not always Books of Logarithms about them, when they have occasion to extract the Square Root: However I will briefly shew you how to do it, and give you one Example thereof.

When you have any Number given whose Square Root you desire, seek for the given Number in the Table of Logarithms under the Title Numbers, and right against it, under the Title Logarithms, you will find the Logarithm of the said Number, the half of which is the Loga-

B 4 rithm

rithm of the Root defired: Which half feek for under the Title Logarithm, and right againſt it under the Title Number, you will find the Root.

EXAMPLE.

Let 625 be the Number whoſe Root is defired; Firſt I feek for it under the Title Numbers, and right againſt it I find this   *Log.* 2, 795889, which I divide by 2, or take the half of it as you ſee: *Half.* 1, 397640, And finding that half under the Title Log. right againſt it is 25, the Root defired. See the ſame done by the former way with leſs trouble.

$$22(0$$
$$625(25 \text{ Root}$$
$$245$$

CHAP.

CHAP. II.

Geometrical DEFINITIONS.

A Point is that which hath neither Length nor Breadth, the least thing which can be imagined, and which cannot be divided, commonly marked as a full stop in Writings thus (.)

A Line has length, but no Breadth nor thickness, and is made by many Points joined together in length, of which there are two sorts, viz. Streight and Crooked. As, A B is a Streight Line, B C two Crooked Lines.

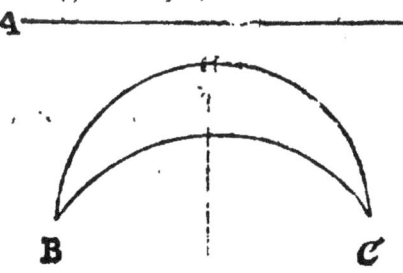

An Angle is the meeting of two Lines in a Point; provided the two Lines so meeting, do not make one Streight Line, as the Line A B, and the Line A C, meeting together in the Point A, make the Angle B A C.

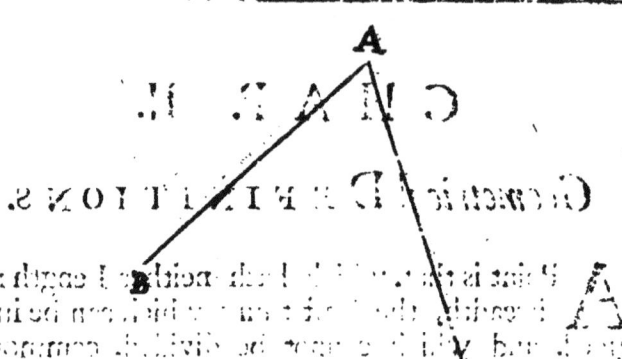

Of which Right-lined Angles there are three sorts, viz. Right-Angled, Acute, Obtuse.

When a Line falleth perpendicularly upon another Line, it maketh two Right Angles.

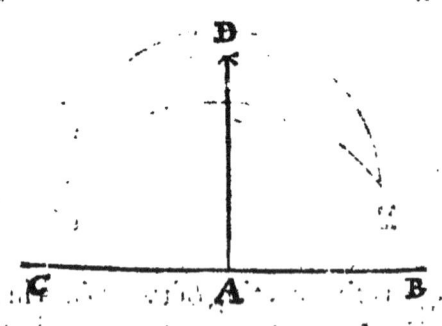

EXAMPLE.

Let C A B be a Right Line, D A a Line Perpendicular to it, that is to say, neither leaning towards B nor C, but exactly upright; then are both the Angles at A, viz. D A B, and D A C, Right Angles; and
contain

Geometrical Definitions.

contain each just 90 Degrees, or the fourth part of a Circle; but if the Line DA had not been Perpendicular, but had leaned towards B, then had D A C been an Obtuse Angle, or greater than a Right Angle, and D A B an Acute Angle, or lesser than a Right Angle, as you see hereunder.

All Figures contained under three Sides are called Triangles, as A, B, C.

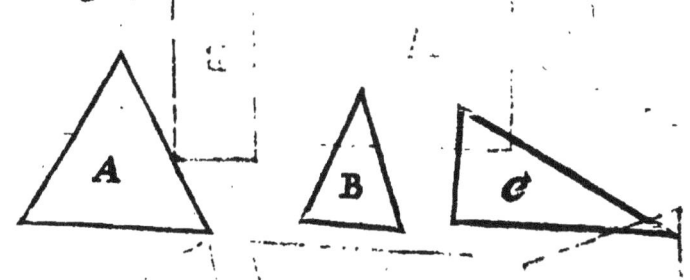

Where note, The Triangle A hath three equal sides, and is called an Equilateral Triangle.

The Triangle B hath two Sides equal, and the third unequal, and is called an Isosceles Triangle.

The Triangle C hath three unequal Sides, and is called a Scalenum.

Of

Geometrical Definitios.

Of four Sided Figures there are these Sorts:

First, a Square, whose Sides are all equal, and Angles Right, as A.

Secondly, A Long Square, or Parallelogram, whose Opposite Sides are equal, and Angles Right, as B.

Thirdly, A Rhombus, whose Sides are all Equal, but no Angle Right, as C.

Fourthly, A Rhomboides, whose Opposite Sides only are Equal, and no Right Angles, as D.

All other four Sided Figures are called Trapezia, as E.

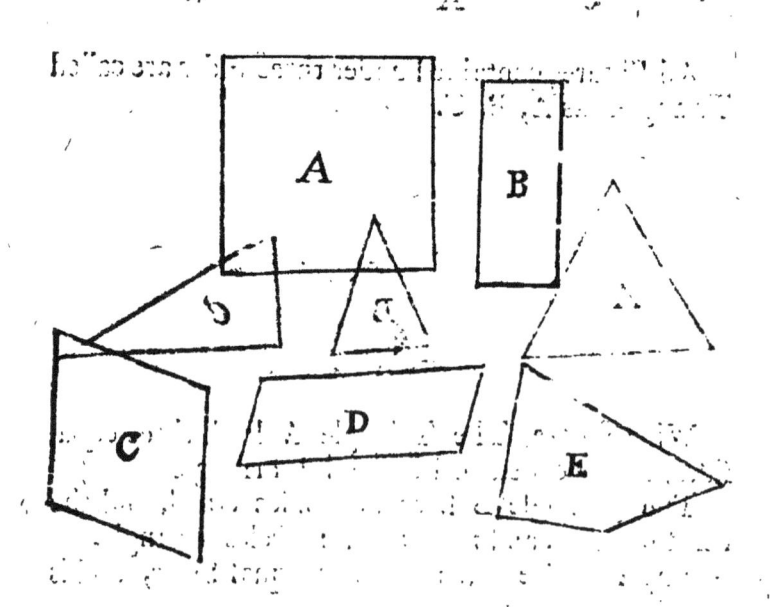

Other

Other Figures that are contained under 5, 6, 7, or more Sides, I call Irregular, as F, &c. Except such as are made by dividing the Circumference of a Circle into any number of equal Parts; for then they are Regular Figures, having all their Sides and Angles Equal; and are called according to the number of Right Lines the Circle is divided into, or more properly according to the Number of Angles they contain, as a Pentagon, Hexagon, Heptagon, Octogon, &c. Which in plain English is no more than a Figure of Five, or Six, Seven or Eight Angles; which Angles are all equal one to another, and their Sides consequently all of the same length. And thus (though I mention no more than 8,) the Circumference of the Circle may be divided into as many Parts as you please; and the Regular Figures arising out of such divisions, are called according to the number of Parts the Circle is divided into; see for your better understanding these two or three following.

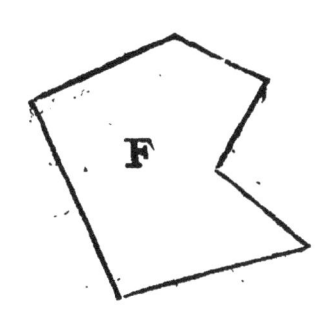

Pentagon 5

Hexagon 6

Heptagon 7

A Circle is a Figure determined with one Endless Line, as A. Which Line is called the Circumference of the Circle, in the Middle whereof is a Prick or Point, by which the Circle is described, which is called the Center, from which Point or Center all Straight Lines drawn to the Circumference are Equal, or of the same Length, as A B, A C, A D.

The

Geometrical Definitions.

The Diameter of a Circle, is a Line which passing through the Center, cuts the Circle into two equal Parts, or the longest Streight Line that can be made in any Circle; as B C.

The Semi-Diameter, is the half of the above-mentioned Line, as A B, A C, or A D, either of which is called a Semi-Diameter.

A Chord, is any Line shorter than the Diameter, which passeth from one part of the Circumference to another, as E C.

A Semicircle is the half of a Circle, as B D C, or B E C.

A Quadrant is the fourth part of a Circle, made by two Diameters perpendicularly interfecting each other, as A B D, A D C, A B E, A E C, either of which is a Quadrant, or the fourth part of a Circle.

A Section, Segment, or part of a Circle is a piece of the Circle cut off by a Chord Line, and is greater or less than a Semicircle, as EFCG is a Segment of the Circle E B D C G, likewise E B D G F is the greater Segment of the same Circle.

A Superficies is that which hath both length and breadth, but no thickness: whose Bounds are Lines, as A is a Superficies or Plain contained in these Lines B C, D E, B D, C E, which hath length from B to C, and Breadth from B to C, but no Thickness.

When

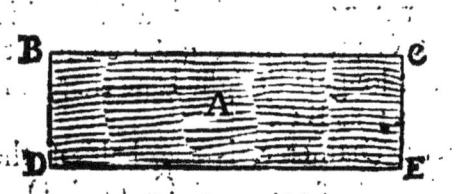

When thefe bounding Lines are meafured, and the Content of the Superficies caft up, the refult is called the Area or Superficial Content of that Figure.

EXAMPLE.

Suppofe the Line BC to be twelve foot in Length, and the Line B D, to be four Foot long, they multiplyed together make 48; therefore I fay 48 Square Feet is the Area or Superficial Content of that Figure.

When two Lines are in every Part equidiftant from each other, they are called Parallel Lines, as the Lines AB and CD, which tho' produced to never fo great a Length, would come no nearer to each other, much lefs meet.

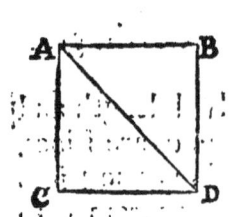

A Diagonal Line is a Line running through a Square Figure, dividing it into two Triangles, beginning at one Angle of the Square and proceeding to the Oppofite Angle. In the Square A B C D, AD is the Diagonal Line.

CHAP.

CHAP. III.

Geometrical PROBLEMS.

PROB. I.

How to make a Line Perpendicular to a Line Given

THE Line given is A B, and at the Point C, it is required to erect a Line which shall be Perpendicular to A B.

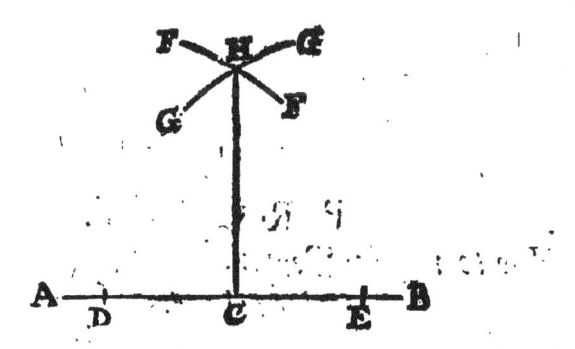

Open your Compasses to any convenient wideness, and setting one Foot of them in the Point C, with the other make a Mark upon the Line at E, and also at D; then taking off your Compasses, open them a little wider than before, and setting one Foot in the Point D, with the other describe the Arch FF, then without altering your Compasses, set one Foot in the Point E, and with the other describe the Arch GG.

C Lastly,

18 *Geometrical Problems.*

Lastly, Lay your Ruler to the Point C, and the Intersection of the two Arches GG, and FF, which is at H, and drawing the Line HC, you have your desire, H C being perpendicular to A B.

See it here done again after the very same manner, but perhaps plainer for your Understanding.

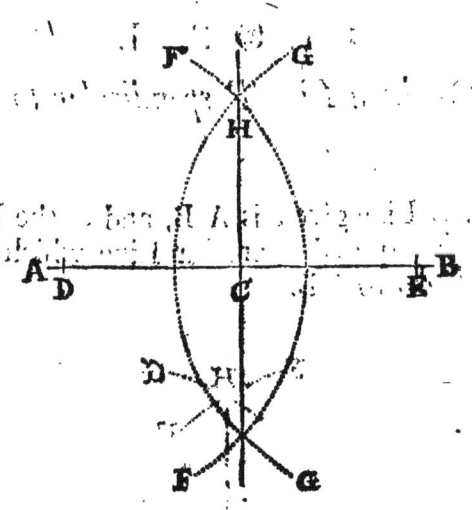

PROB. ii.
How to raise a Perpendicular upon the end of a Line.

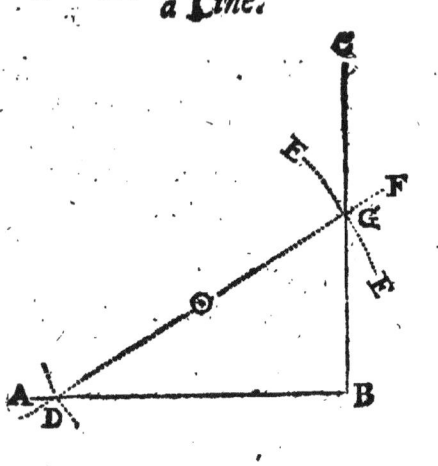

Geometrical Problems

A B is the Line given, and at B it is required to erect the Perpendicular B C.

Open your Compasses to an ordinary extent, and setting one Foot in the Point B, let the other fall at adventure, no matter where in Reason, as at the Point ⊙, then without altering the extent of the Compasses, set one Foot in the Point ⊙, and with the other cross the Line AB as at D: Also on the other side describe the Arch E E, then laying your Ruler to D and ⊙, draw the prickt Line D ⊙ F. Lastly, from the Point B, you began at, through the Intersection at G, draw the Line B G C, which is perpendicular to AB.

Another way, I think more easie, though indeed almost the same.

Let A B be the given Line, B I the Perpendicular required.

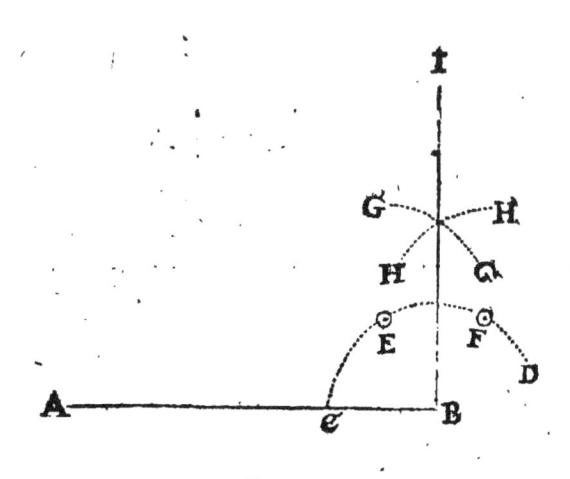

20 Geometrical Problems.

Set one Foot of your Compasses in B, and with the other at any ordinary extent, describe the Arch CEFD, then keeping your Compasses at the same extent, set one Foot in C, and make a Mark upon the Arch at E; and keeping one Foot in E, make another Mark at F, then with any extent, set one Foot in E, and with the other describe the Arch GG: Also setting one Point in F, make the Arch HH, then drawing a Line through the intersection of the Arches G and H, to the Point first proposed

PROB. iii.

How from a Point assigned, to let fall a Perpendicular upon a Line given

The Line given is A B, the Point is at C, from which it is desired to draw a line down to AB, that may be Perpendicular to it;

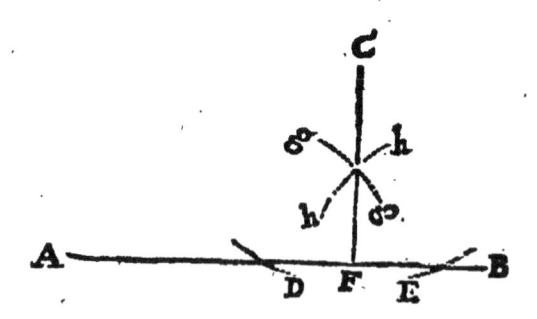

First, setting one Foot of your Compasses in the Point C, with the other make a Mark upon the Line A B as at D, and also at E, then opening your Compasses wider, or shutting them closer, either will do;

set

Geometrical Problems.

set one Foot in the Point of Intersection at D, and with the other describe the Arch *g g*, the like do at E, for the Arch *h h*. Lastly, from the Point assigned, through the Point of Intersectihn of the two Arches, *g g*, and *h h*, draw the Perpendicular Line CF. This is no more but the first Problem reversed: The same you may do by the second Problem, *viz.* Let fall a Perpeudicular nigh the end of a given Line.

PROB. iv.

How to Divide a Line into any Number of Equal Parts.

A B is a Line given, and it is required to divide it into 6 equal parts.

Make at the Point B a Line Perpendicular to A B, as B C: Do the same at A, the contrary way, as you see here, open your Compasses to any convenient Wideness, and upon the Lines B C, and A D, mark out five Equal Parts; for it must be always one less than the Number you intend to divide the Line into: Which Parts you may number, as you see here,

here, thofe upon one Line one way, and the other the contrary way; then laying your Ruler from N°. 1. on the Line B C. to N°. 1. on the Line A D, is will interfect the Line A B at E, which you may mark with your Pen, and the Diftance between B, and E. is one fixth part of the Line; fo proceed on 'till you come to N°. 5. and then you will find that you have divided the given Line into fix Equal Parts, as required.

PROB. v.

How to make an Angle Equal to any other Angle given.

The Angle given is A, and you are defired to make one Equal to it.

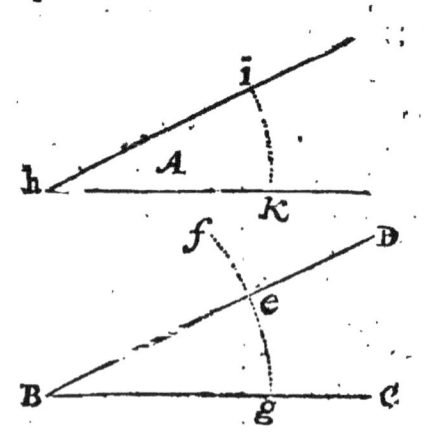

Draw the Right Line BC, then going to the Angle A, fet one Foot of your Compaffes in the Point *h*, and with the other at what Diftance you pleafe, defcribe describe the Arch I K, then without altering the extent of the Compasses, set one Foot in B, and draw the like Arch, as *f g*; after that, measure with your Compasses how far it is from K to I, and the same Distance set down upon the Arch from *g* towards *f*, which will fall at C; after draw the Line B C D, and you have done.

PROB. vi.

How to make Lines Parallel to each other.

A B is a Line given, and it is required to make a Line parallel unto it.

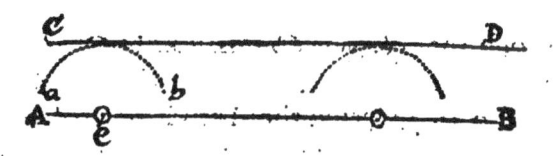

Set one foot of your Compasses at or near the end of your given Line as at C, and with the other describe the Arch *a b*; do the same near the other end of the same Line, and through the utmost Convex of those two Arches draw the line C D. it is the Parallel required.

PROB. vii.

How to make a Line Parallel to another line, which must also pass through a Point assigned.

Let A B, be the given line, C the point through which the required Parallel line must pass.

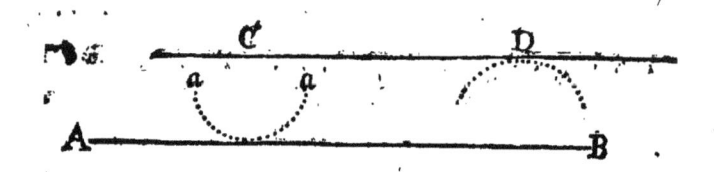

Set one foot of your Compasses in C, and closing them so that they will just touch, (and no more) the Line A B, describe the Arch *a a*; with the same extent in any part of the given Line set one Foot, and describe another Arch as at D: then through the assigned Point C, and the utmost Convex of the last Arch, draw the required Line C D, which is Parallel to A B, and passeth through the Point C.

PROB. viii.

How to make a Triangle, three Lines being given

Let the three lines given be 1, 2, 3, The Question is how to make a Triangle of them.

Take

Take with your Compasses the length of either of the three in this Example; let it be that N°. 1. *viz.* the longest, and lay it down as hereunder from A to B; then taking with your Compasses the length of the Line 2, set one Foot in B, and make the Arch C; also taking the length of the last Line 3 place your Compasses at A, and make the Arch D, which will intersect the Arch C, at the Point e; from which Point of Intersection draw Lines to A B, which shall constitute the Triangle A e B; The Line A B being equal to the line N°. 1, B e to N°. 2, A e to N°. 3.

PROB. ix.

How to make a Triangle equal to a Triangle given, and every way in the same Proportion.

First make an Angle Equal to the Angle at A, as you were taught in PROB. v. Then making the Lines AD and AE equal to A B and A C, draw the Line D E.

Or otherwise you may do it as you were taught in PROB. viii.

PROB.

PROB. x.

How to make a Square Figure.

Let A be a Line given, and it is required to make a square Figure, each side of which shall just be the length of the Line A.

First lay down the length of your Line A, as A B.

Secondly, raise a Perpendicular of the same length at B.

Thirdly, take the length of either of the aforementioned Lines with your Compasses, and setting one Foot in C describe the Arch *e e*; do the like at A, and describe the Arch *f f*.

Fourthly, draw Lines from A and C into the Point of Intersection, and the Square is finished.

PROB. xi.

How to make a Parallelogram, or long Square.

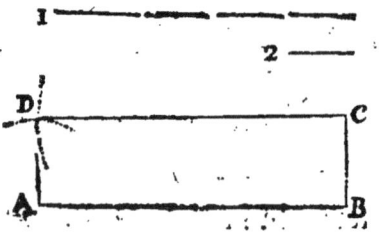

This is much like the former. Admit two Lines be given as 1, 2, and it is required to make a Parallelogram of them: What a Parallelogram is, you may see in the Second Chapter of *Definitions*.

First, lay down your longest Line, as A B, upon the end of which erect a Perpendicular Line, equal in Length to your shortest Line, and so proceed, as you were taught in the foregoing Problem.

PROB. xii.

How to make a Rhombus

First make an Angle, suppose A C B, no matter how great or small; but be sure let the two Lines be of equal length; then taking with your Compasses the length of one of those two Lines, set one Foot in A, and describe the Arch *b b*, also set one Foot in B, and describe the Arch *c c*. Lastly, draw Lines, and it is finished. Two Equilateral Triangles is a Rhombus.

A Rhomboides differs just so much, and no more from a Rhombus, as a Parallelogram does from a true Square; it is needless therefore, I presume, to shew you how to make it.

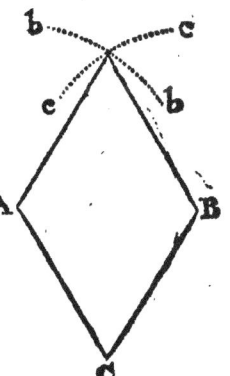

PROB.

PROB. xiii.

How to divide a Circle into any number of Equal Parts, not exceeding ten, or otherwise how to make the Figures called, Pentagon, Hexagon, Heptagon, Octagon, &c.

Let A B C D be a Circle, in which is required to be made a Triangle, the greatest that can be made in that Circle.

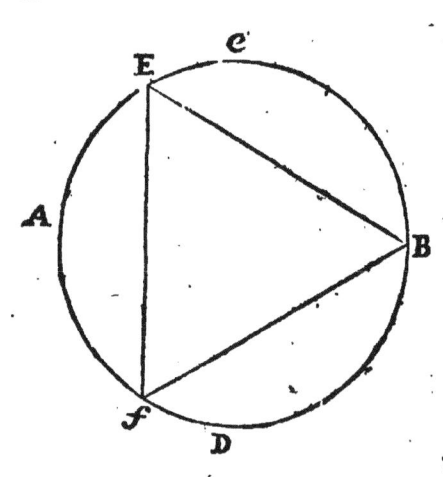

Keeping your Compasses at the same extent they were at when you made the Circle, set one Point of them in any part of the Circle, as at A, and with the other make a Mark at E and *f*, and draw a Line between E and *f*, which will be one side of the Triangle.

I need not tell you how to make the other two Side, for it is an Equilateral Triangle, all three Sides being of Equal Length.

To make a Pentagon or Five-sided Figures

Draw first an obscure Circle, as A C B D; then draw a Diameter from A to B; make another Diameter Perpendicular to the first, as CD; then taking with your Compasses the Length of the Semi-Diameter, set one Point in A, and make the Marks B F, drawing a Line between them, as you did to make Triangle. Next, set one Point of your Compasses in the Intersection at g, and extend the other to C, draw the Arch C H: The nearest distance between C and H, viz. the Line C I H, is the Side of a Pentagon, and the greatest that can be made within that Circle: Which with the same extent of your Compasses, you may mark out round the Circle, and drawing Lines, the Figure will be finished.

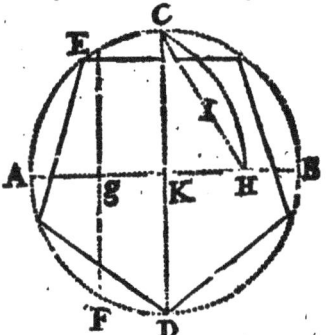

To make a Hexagon or Six-sided Figure.

Draw an obscure Circle, as you see here, and then without altering the extent of the Compasses, mark out the Hexagon required round the Circle; for the Semidiameter of any Circle is the side of the greatest Hexagon that can be made within the same Circle.

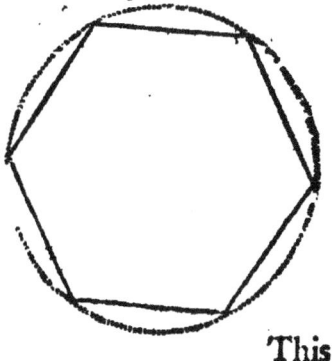

This

30 Geometrical Problems.

This is the way Coopers use to make Heads for their Casks.

To make a Heptagon, or Figure of Seven, equal Sides and Angles.

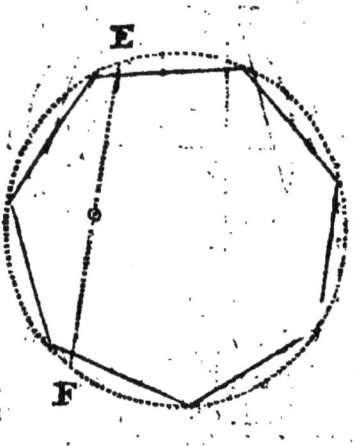

You must begin and proceed as if you were going to describe a Triangle in a Circle, till you have drawn the Line E F; then taking with your Compasses the half of that Line, viz. from ⊙ to E, or from ⊙ to F, mark out round the Circle your Heptagon for the half of the Line E F is one side of it.

To make an Octogon, commonly called an Eight-square Figure.

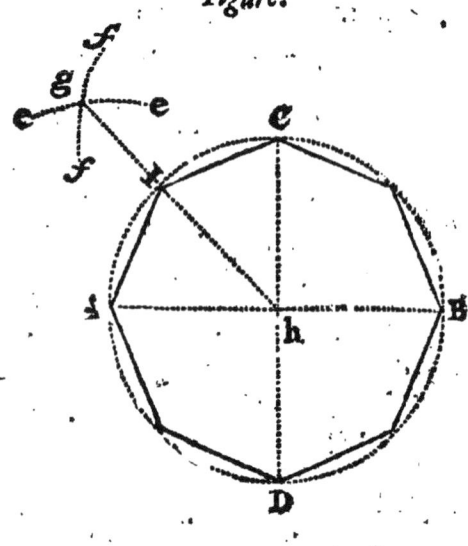

First

First make a Circle.

Secondly, Divide it into four equal Parts by two Diameters, the one perpendicular to the other, as A B and C D.

Thirdly, Set one Foot of the Compasses in A, and make the Arch *e e*, also with the same extent set one foot C, and make the Arch *f f*; then through the Intersection of the two Arches draw a Line to the Center, viz. *g h*.

Lastly, Draw the Line I C or I A, either of which is the side of an Octagon.

To make a Nonagon

First make a Circle, and a Triangle in it, as you were taught at the beginning of this Problem. Then divide one third part of the Circle. As for Example, that A, 1, 2, 3, B, into three equal Parts. Lastly, draw the lines A 1, 1, 2, 2 B, &c. each of these Lines is the side of a *Nonagon*.

To make a Decagon.

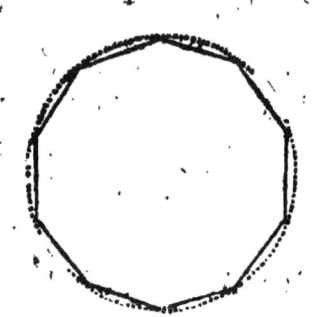

You must work altogether as you did in making a *Pentagon*: See the *Pentagon* above, where the distance from the Centre K to the Point at H is the side of a *Decagon* or Ten-sided Figure.

PROB. xiv.

Three Points being given: How to make a Circle, whose Circumference shall pass through the three given Points, provided the three Points are not in a streight Line.

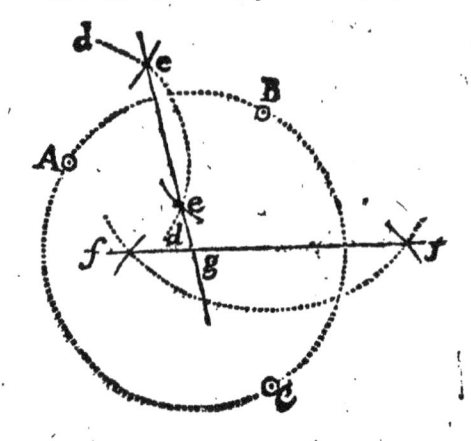

Let A, B, C, be the three Points given; first setting one Foot of your Compasses in A, open them to any convenient wideness, more than half the distance between A and B, and describe the Arch *d d*; then without altering the extent, set one point in B, and cross the first Arch at *e* and *e*, through those two Intersections draw the Line *e e*.

The very same you must do between B and C, and draw the Line *ff*; where those two Lines intersect each other, as at *g*, there is the Centre of the Circle required; therefore setting one foot of your Compasses in *g*, extend the other to any of the Points given, and describe the Circle A B C. Note, The Centre of a Triangle is found the same way.

PROB. XV.
How to make an Ellipsis, or Oval, several ways.

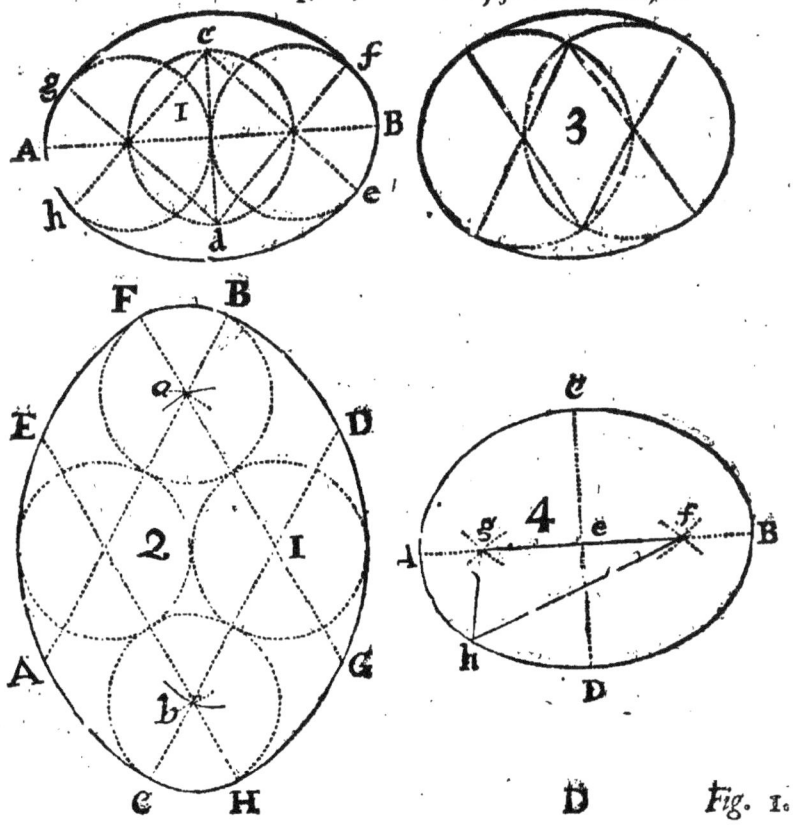

Fig. 1.

Fig. 1. Make the Circles whose Diameters may be in a streight Line, as A B; Cross that Line with another Perpendicular to it at the Centre of the middle Circle, as *c d*: draw the Lines *c e*, *c h*, *d g*, *d f*. Set one foot of the Compasses in D, and extend the other to *g*, describing the part of the Ellipsis *g f*, with the same extent, setting one foot in C, describe the other part *h e*: The two Ends are made by parts of the two outermost small Circles, as you see *f e*, *g h*.

Fig. 2. Draw two small Circles, whose circumferences may only touch each other: Then taking the distance between their Centers, or either of their Diameters, set one foot of your Compasses in either of their Centres, as that marked 2, and with the other make an Arch at *a*, also at *b*; there moving your Compasses to the Centre of the other Circle, cross the said Arches at *a* and *b*, which Crosses let be the Centres of two other Circles of equal bigness with the first. Then through the Centres of all the Circles, draw the lines A B, C D, E H, F G; which done, place one foot of the Compasses in the centre of the Circle I, and extend the other to C, describing the Arch of the Ellipsis CEF: The same you must do at 2, to describe the part BH, and then is your Ellipsis finished.

Fig. 3. This needs no Description, it being so like the two former Figures, and easier than either of them.

Here Note, that you may make the Ovals 1 and 3 of any determined length; for in the length of the first, there is four Semi-diameters, of the small Circles; and in the last but three: If therefore any line was given you, of which length an Oval was required, you must take in your Compasses the

fourth

fourth part of the line, to make the Oval *Fig.* 1. and the third part to make the Oval *Fig.* 3; and with that extent you muſt deſcribe the ſmall Circles: The Breadth will be always proportional the length. But if the Breadth be given you, take in alſo the fourth part thereof, and make the Oval *Fig.* 2.

Fig. 4. This Ellipſis is to be made, having Length and Breadth both given. Let A B be the length, CD the breadth of a required Oval. Firſt lay down the line A B equal to the given length, and croſs it in the middle with the Perpendicular CD, equal to the given breadth. Secondly, Take in half the Line A B with your Compaſſes, *viz.* A e, or B e; ſet one foot in C, and make two marks upon the line AB, *viz.* ƒ and *g*; alſo with the ſame extent ſet one foot in D, and croſs the former marks at ƒ and *g*. Thirdly, at the Point ƒ and *g*, fix two Pins; or if it be a Garden-plat, or the like, two ſtrong Sticks. Then putting a line about them, make faſt the two ends at ſuch an exact length, that ſtretching by the two Pins, the bent of the line may exactly touch A or B, or C or D, or *h*, as in this Diagram it does at *h*; ſo moving the line ſtill round, it will deſcribe an exact Oval.

D 2   P R O B.

PROB. xvi.

How to divide a given Line into two Parts, which may be in such Proportion to each other, as two given Lines.

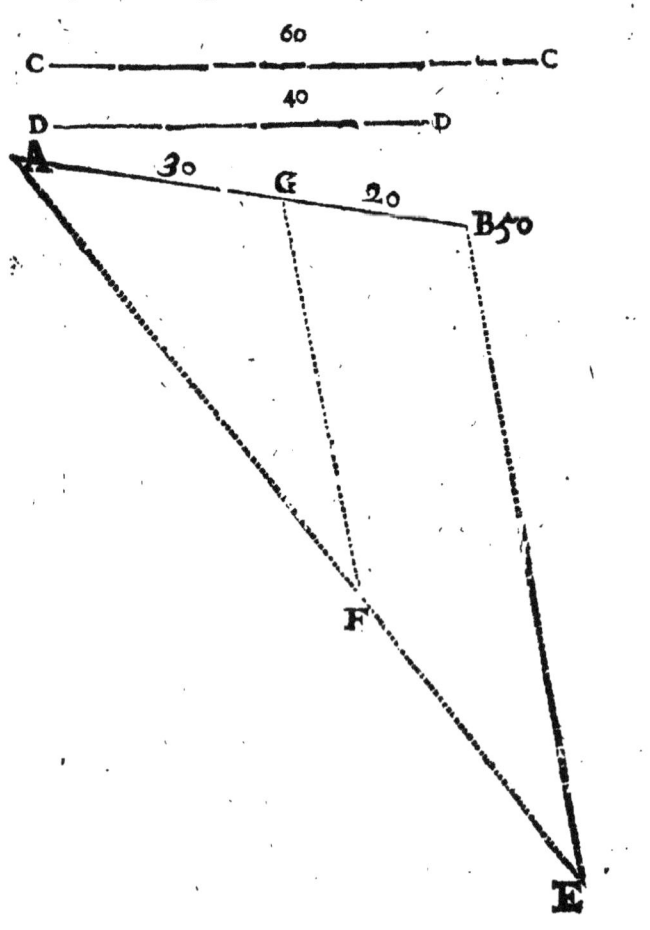

Let

Let AB be the given line to be divided in such Proportion as the line C is to the line D.

First from A draw a line at pleasure, as A E; then taking with your Compasses the line C, set it off from A towards E, which will fall at F: Also take the line D, and set off from F to E.

Secondly, Draw the Line EB; and from F make a line Parallel to E B, as FG, which shall intersect the given line A B in the proportional Point required, *viz.* at G; making A G and G B in like proportion to each other, as C C and D D.

Example by Arithmetick.

The line C C, is 60 Feet, Perches, or any thing else; the line D D is 40; the line A B is 50; which is required to be divided in such proportion as 60 to 40. First add the two lines C and D together, and they make a 100: Then say, If 100 the whole give 60 for its greatest part, what shall 50, the whole line A B, give for its greatest proportional part? Multiply 50 by 60, it makes 3000; which divided by 100, produces 30 for the longest part; which 30 taken from 50, leaves 20 for the shortest part: as therefore 60 is to 40, so is 30 to 20.

PROB. xvii.

Three Lines being given, to find a Fourth in Proportion to them.

Let A B C be the three Lines given, and it is required to find a fourth Line which may be in such proportion to C, as B is to A,

A———14
B———18
C————21

which

Geometrical Problems.

which is no more but performing the *Rule of Three* in Lines. As if we should say, if A 14 give B 18, what shall C 21 give? Answer, 27. But to perform the same Geometrically, work thus.

First make any Angle, as B A C. Then take with your Compasses the first line, A and set it from A to 14. Also take the second line B, and set it from A to 18; draw the line 14, 18. Then take the third line C with your Compasses, and set it from A to 21. From 21 draw a line parallel to 14, 18, which will be 21, 27. Then from A to 27, is the length of your Fourth Line required.

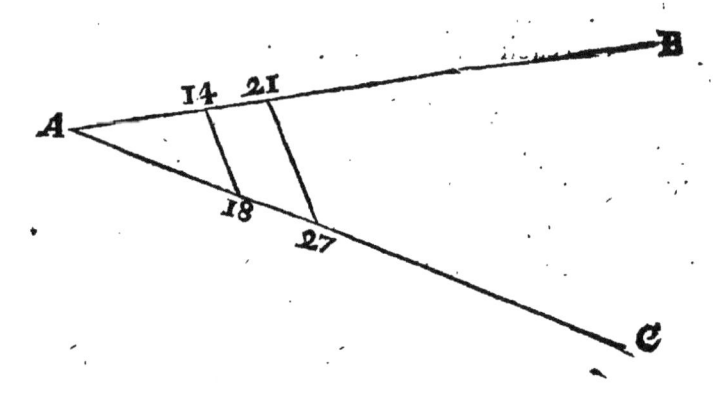

And here for a while I shall leave these *Problems*, till I come to shew you how to divide any piece of Land; and to lay out any piece of a given quantity of Acres into any Form or Figure required: And in the mean time I shall shew you what is necessary to be known.

CHAP.

CHAP. II.

Of Measures.

AND first of Long Measures; which is either Inches, Feet, Yards, Perches, Chains, &c. Note that twelve Inches make one Foot, three Feet one Yard, five Yards and an half one Pole or Perch, four Perches one Chain of *Gunter's*, eighty Chains one Mile. But if you would bring one sort of Measure into another, you must work by *Multiplication* or *Division*. As for Example, Suppose you would know how many Inches are contained in twenty Yards: First reduce the Yards into Feet, by multiplying them by 3, because 3 Feet make one Yard, the Product is 60, which multiplied by 12, the number of Inches in one Foot, gives 720, and so many Inches are contained in 20 Yards length.

On the contrary, if you would have known how many Yards there are in 720 Inches, you must first divide 720 by 12, the Quotient is 60 Feet; that again divided by 3, the Quotient is 20 Yards. The like you must do with any other Measure, as Perches, Chains, &c. of which more by and by.

Long	Link	Foot	Yard	Perch	Chain	Mile
Inches	7.92	12	36	198	792	63360
	Links	1.515	4.56	25	100	8000
		Feet	3	16.5	66	5280
			Yard	5.5	22	1760
				Perch	4	320
					Chain	80

See this Table of the Long. Meafure annexed, the ufe whereof is very eafie: If you would know how many Feet in Lengh, go to make one Chain; look for Chain at the Top, and at the Left-hand for Feet, againft which, in the common Angle of Meeting, is 66, fo many Feet are contained in one Chain.

But becaufe Mr. *Gunter*'s Chain is moft in ufe amang Surveyors for meafuring of Lines, I fhall chiefly infift on that meafure, it being the beft in ufe for Lands,

This Chain contains in Length 4 Pole, or 66 Feet, and is divided into 100 Links, each Link is therefore in length 71$\frac{51}{100}$ Inches: If you would turn any number of Chains into Feet, you muft multiply them by 66, as 100 Chains multiplied by 66, makes 6600 Feet; but if you have Links to your Chains to be turned into Feet and parts of Feet, you muft fet down the Chains and Links, as if they were one whole Number, and after having multiplied that Number by 66, cut off from the Product the two laft Figures to the Right-hand, which will be the Hundreth Parts of a Foot, and thofe on the Left-hand the Feet required.

EXAMPLE.

Let it be required to know how many Feet there are in 15 Chains, 25 Links.

I fet down thus the *Multiplicand*   1525
The num. of Feet in 1 Chain, *Multiplicat.* 66

            9150
            9150

Product 1006|50 Feet

Of Measure

The *Product* is $1006\frac{50}{100}$. This is so plain, it needs no other Example.

But now on the other hand, if One thousand and six Feet and an halfe was given you to reduce into Chains and Links; you must divide 100650 by 66, the Quotient will be 1525, *viz.* 15 Chains, 25 Links. But for those that do not well understand *Decimal Arithmetick*, and may perhaps meet with harder Questions of this nature, I have inserted this Table.

A Table, shewing how many Feet and Parts of a Foot; also how many Perches and Parts of a Perch, are contained in any number of Chains and Links, from One Link, to One hundred Chains.

Links	Feet	Parts of a Foot	Perches	Part of a Perch	Chains	Feet	Perches
1	00	66	0	04	1	66	4
2	01	32	0	08	2	132	8
3	01	98	0	12	3	198	12
4	02	64	0	16	4	264	16
5	03	30	0	20	5	330	20
6	03	96	0	24	6	396	24
7	04	62	0	28	7	462	28
8	05	28	0	32	8	528	32
9	05	94	0	36	9	594	36
10	06	60	0	40	10	660	40
20	13	20	0	80	20	1320	80
30	19	80	1	20	30	1980	120
40	26	40	1	60	40	2640	160
50	33	00	2	00	50	3300	200
60	39	60	2	40	60	3960	240
70	46	20	2	80	70	4620	280
80	52	80	3	20	80	5280	320
90	59	40	3	60	90	5940	360
100	66	00	4	00	100	6600	400

The Explanation of the Table.

If you would know how many Feet are contained in Twenty of Mr. *Gunter*'s Chains.

First, under Title *Chains*, seek for 20; and right against it, under Title *Feet*, stands 1320, the number of Feet contained in Twenty Chains. Also under Title *Perches*, stands 80, the number of Perches contained in Twenty Chains.

Again

Of Measure.

A Table, shewing how many Chains, Links, and Parts of a Link; also how many Perches and parts of a Perch, are contained in any number of Feet, from 1 to 10000.

Feet	Chain	Link	P.of L.	Perch	P.of Per
1	0	1	515	0	060
2	0	3	030	0	121
3	0	4	545	0	181
4	0	6	060	0	242
5	0	7	575	0	303
6	0	9	090	0	363
7	0	10	606	0	424
8	0	12	121	0	484
9	0	13	636	0	545
10	0	15	151	0	606
20	0	30	303	1	212
30	0	45	454	1	818
40	0	60	606	2	424
50	0	75	757	2	030
60	0	90	909	3	636
70	1	06	060	4	242
80	1	21	212	4	848
90	1	36	363	5	454
100	1	51	515	6	060
200	3	03	030	12	121
300	4	54	545	18	181
400	6	06	060	24	242
500	7	57	575	30	303
600	9	09	090	36	363
700	10	60	606	42	424
800	12	12	121	48	484
900	13	63	636	54	545
1000	15	15	151	60	606
2000	30	30	303	121	212
3000	45	45	454	181	818
4000	60	60	606	242	424
5000	75	75	757	303	030
6000	90	90	909	363	636
7000	106	06	060	424	242
8000	121	21	212	484	848
9000	136	36	363	545	454
10000	151	51	515	606	060

Of Measure

This Table is like the former, and needs not much Explanation. However I will give an Example or two.

Admit I would know how many Chains in length are contained in 500 Feet. First, in the left-hand Column, under Title *Feet*, I look out 500, and right against it I find 7 Chains, 57 Links, 575 Parts of 1000 of a Link, or 7 Chains, 57 $\frac{575}{1000}$ Links. So likewise under Title *Perches*, I find 30 $\frac{303}{1000}$ Perches. But if you would know how many odd Feet that $\frac{303}{1000}$ is, you must seek for 303 in the Column titled *Parts of a Perch*, and right against it you will find 5 Feet. So I say that 500 Feet is 30 Perches 5 Foot.

Again, I would know how many Chains and Links there are in 15045 Feet; First seek for 10000, and write down the Chains, Links, and part of a Link contained therein. Do the like by 5000; also by 40 and 5. Lastly, adding them together, you have your desire.

Feet,	Chain,	Link,	Parts
10000 —	151	51	515
5000 —	75	75	757
40 —	0	60	606
5 —	0	7	575
Added, make —	227	95	453

Answer, 227 Chains, 95 Links and 453 parts are contained in 15045 Feet.

One *Example* more, and I have done with this Table.

How many Perches do 10573 Feet make?

Feet,

Feet,	Perches,	Parts.
10000	606	060
500	30	303
70	4	242
3	0	181
Add 640		786

The Answer is, 640 Perches, and $\frac{786}{1000}$ of a Perch, or 13 Feet, a Furlong is 40 Perches in length; 8 Furlongs make 1 Mile. And so much of *Long Measure*: I shall now proceed to.

Square Measure.

Planometry, or the measuring the Superfices or Planes of things (as Sir *Jonas Moore* says) is done with the Squares of such Measures, as a Square Foot, a Square Perch, or Chain, that is to say, by Squares whose Sides are a Foot, a Perch, or Chain; and the Content of any Superficies is said to be found, when we know how many such Squares it containeth.

As for Example: Suppose A B C D was a piece of Land, and the length of the Line A B or C D was 4 Perches; also the length of the Line A C or B D was 5 Perches; I say that piece of Land contains 20 Square Perches, as you may see it here divided; every little Square being a Perch, having a Perch in length for its Side. If you lay down a Square Figure, whose Side is 1 Foot, and

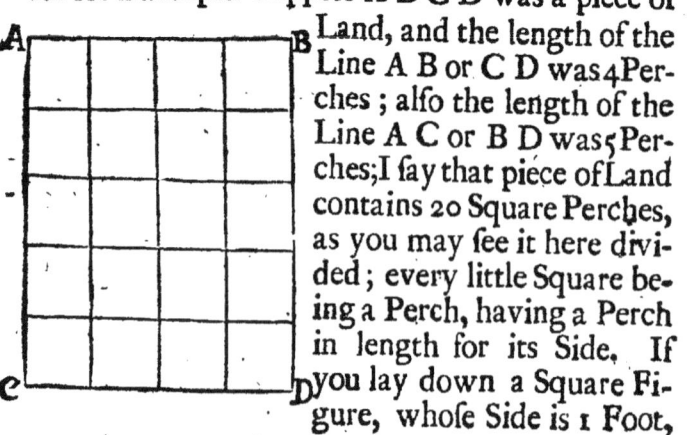

Geometrical Problems

and at the end of every Inch you draw Lines crossing one another, as these here, you will divide that Square Foot into 144 little Squares, or square Inches.

Or thus, the Line *a b* is a Perch long or 16 Feet ½, so is the Line *b d*, and the other 2 Lines: The whole Figure *a b c d* is called a Square Perch.

But before we go any farther, take this Table following of Square Measure.

A TABLE of SQUARE MEASURE.

	Inch	Links	Feet	Yards	Pace	Perch	Chain	Acre	Mile
Inch	1								
Links	62.726	1							
Feet	144	2.295	1						
Yards	1296	20.755	9	1					
Pace	3600	57.381	25	2.778	1				
Perch	39204	625	272.25	30.25	10.89	1			
Chain	627264	10000	4356	484	174.24	16	1		
Acre	6272640	100000	43560	4840	1742.4	160	10	1	
Mile	4014489600	64000000	27878400	3097600	1115136	102400	6400	640	1

This Table is like the former of Long Measure, and the use of is is the same.

Example, If you would know how many Square Feet are contained in one Chain, look for Feet at Top, and Chain on the Side, and in the common Angle of meeting stands 4356, so many Square Feet are contained in one Square Chain.

The common Measure for Land is the Acre, which by Statute is appointed to contain 160 Square Perches, and it matters not in what form the Acre lye, so it contains just 160 Square Perches; as in a Parallelogram 10 Perches one way, and 16 another, contain an Acre: So does 8 one way and 20 another, and 4 one way, and 40 another. If then, having one Side given in Perches, you would know how far you must go on the Perpendicular to cut of an Acre? you must divide 160 (the number of Spuare Perches in an Acre) by the given Side, the Quotient is your desire. As for *Example*, the given Side is 20 Perches, divide 160 by 20, the Quotient is 8: By that I know, That 20 Perches one way, and 8 another, including a Right Angle, will be the two Sides of an Acre; the other two Sides must be parallel to these.

And here I think it convenient to insert this necessary Table, shewing the length and breadth of and Acre in Perches, Feet and Parts of a Foot: But if your given Side had been in any other sort of Measure; As For Instance in Yards, you must then have seen how many Square Yards had been in a Acre, and that Summ you must have divided by the number of your given Yards, the Quotient would have answered the Question.

EXAMPLE.

Of Measure 49

EXAMPLE.

If 44 Yards be given for the Breadth, how many Yards shall there be in length of the Acre?

First, I find that an Acre contains 4840 Square Yards, which I divid by 44, the Quotient is 110 for the Length of an Acre. And thus knowing well how to take the Length and Breadth of an Acre, you may also by the same way know how to lay down any number of Acres together; of which more hereafter.

Reducing of one sort of Square Measure to another, is done, as before taught in Long Measure, by Multiplication and Division. And because Mr. *Gunter's* Chain is chiefly used by Surveyors, I shall only Instance in that, and shew you how to turn any number of Chains and Links into Acres, Roods and Perches: Note that a Rood is the fourth part of an Acre.

And first mark well that 10 square Chains make 1 Acre,

Breadth	Length of an Acre		Breadth	Length of an Acre	
Pearches	Perches	Feet	Perches	Perches	Feet
10	16	0	28	5	11¼
11	14	9	29	5	8½
12	13	5½	30	5	5½
13	12	5⅓	31	5	2⅔
14	11	7½	32	5	0
15	10	11	33	4	14
16	10	0	34	4	11½
17	9	6½	35	4	9
18	8	14½	36	4	7½
19	8	6½	37	4	5¼
20	8	0	38	4	3½
21	7	10¾	39	4	1⅓
22	7	4¼	40	4	0
23	6	15¼	41	3	14½
24	6	11	42	3	13½
25	6	6½	43	3	11½
26	6	2½	44	3	10½
27	5	15½	45	3	2¼

Of Measure.

1 Acre, that is to say, 1 Chain in Breadth, and 10 in Length; or 2 in Breadth, and 5 in Length, is an Acre, as you may see by this small Table.

Chains (Length of an Acre)	Chains (Breadth of an Acre)	Links	Parts of a Link
1	10	00	
2	5	00	
3	3	33	333
4	2	50	
5	2	00	
6	1	66	666
7	1	42	285
8	1	25	
9	1	11	111

And thus well weighing that 10 Chains make an Acre, if any number of Chains be given you to turn into Acres, you must divide them by 10, and the Quotient will be the number of Acres contained in so many Chains. But this Division is abbreviated by only cutting off the last Figure, as if 1590 Chains were given to turn into Acres, by cutting off the last Figure 159|0, there is left 159 Acres, which is all one as if you had divided 1590 by 10. But if Chains and Links be given you together to turn into Acres, Roods and Perches, first from the given Summ cut off three Figures, which is two Figures for the Links and one for the Chains, what's left shall be Acres. And to know how many Roods and Perches are contained in the Figures cut off, multiply them by 4, and from the Product, cutting off the three last Figures, you will have the Roods: And then to know the Perches, multiply the Figures cut off from the Roods, by 40, from which Product cutting off again three Figures, you have the Perches, and the Figures cut off are thousandth parts of a Perch.

EXAMPLE

Of Measure.

EXAMPLE.

In 1599 Square Chains, and 55 Square Links, how many Acres, Roods and Perches?

Answer, 159 Acres, 3 Rood 32 $\frac{8}{100}$

Acres   159'955
            4
           ———
Roods   3'620
           40
           ———
Perches 24|800

On the contrary, if to any number of Acres given, you add a Cypher, they well be turned into Chains, thus 99 Acres are 990 Chains, 100 Acres 1000 Chains, &c. the same as if you had multiplied the Acres by 10. And if you would turn Square Chains into Square Links, add four Cyphers to the end of the Chains, so will 990 Chains be 9900000 Links, 1000 Chains be 10000000 Links, all one as if you had multiplied 990 by 10000, the number of Square Links contained in one Chain.

And now, whereas in casting up the content of a piece of Land measured by Mr. *Gunter's* Chain, (viz. multiplying Chains and Links by Chains and Links) the Product will be Square Links; you must therefore from that Product cut off five Figures to find the Acres; which is the same as if you divide the Product by 100000 (the number of Square Links contained in one Acre) then multiply the five Figures cut off by 4; and from that Product cutting off five Figures you will have the Roods. Lastly multiply by

E 2   40,

40, and take away (as before) 5 Figures, the ref are Perches.

EXAMPLE.

Admit a Parallelogram, or Long Square to be one way 5 Chains, 55 Links; and the other way 4 Chains, 35 Links: I demand the content in Acres, Roods and Perches?

Multiplicand 555
Multiplicator 435

2775
1665
2220

Answer, 2 Acres }
1 Rood }
26 Perches }
And 2¾ Parts of a Perch.

Acres 2|4142
4
Roods 1|6570
40
Perches 26|2800

Lastly, Because some Men chuse rather to cast up the Content of Land in Perches, I will here briefly shew you how it is done; which is, only by dividing by 160 (the number of Square Perches contained in one Acre) the number of Perches given.

EXAMPLE.

Admit a Parallelogram to be in length 55 Perches, and in breadth 45 Perches; these two multiplied together, make 2475 Perches; which to turn into Acres, divide by 160, the Quotient is 15 Acres, and 75 Perches remaining; which to turn into Roods divide by 40, the Quotient is 1 Rood, and 35 Perches remaining. So much is the content of such a piece of Land, viz. 15 Acres, 1 Rood, and 35 Perches.

Of Measure

Here follows a Table to turn Perches into Acres, Roods and Perches.

Perches	Acres	Roods	Perch
40	0	1	00
50	0	1	10
60	0	1	20
70	0	1	30
80	0	2	00
90	0	2	10
100	0	2	20
200	1	1	00
300	1	3	20
400	2	2	00
500	3	0	20
600	3	3	00
700	4	1	20
800	5	0	00
900	5	2	20
1000	6	1	00
2000	12	2	00
3000	18	3	00
4000	25	0	00
5000	31	1	00
6000	37	2	00
7000	43	3	00
8000	50	0	00
9000	56	1	00
10000	62	2	00
20000	125	0	00
30000	187	2	00
40000	250	0	00
50000	312	2	00
60000	375	0	00
70000	437	2	00
80000	500	0	00
90000	562	2	00
100000	625	0	00

The Use of the Table.

In 2475 Perches, how many Acres, Roods and Perches.

Perch	Acres	Rood	Perch
2000	12	2	00
400	2	2	00
70	0	1	30
To which add the odd 5 Perches	0	0	5
Answer	15	1	35

CHAP.

CHAP. V.

Of Inftruments and their Ufes.

And firft of the Chain.

THere are feveral forts of Chains, as Mr. *Rathborne*'s of two Perch long: Others, of one Perch long: fome have had them 100 Feet in length. But that which is moft in ufe among Surveyors (as being indeed the beft) is Mr. *Gunter*'s, which is 4 Pole long, containing 100 Links, each Link being $7\frac{92}{100}$ Inches: The Defcription of which Chain, and how to reduce it into any other Meafure, you have at large in the foregoing Chapter of Meafure. In this place I fhall only give you fome few Directions for the ufe of it in meafuring Lines.

Take care that they who carry the Chain, deviate not from a ftreight Line; which you may do by ftanding at your Inftrument, and looking thro' the Sights: If you fee them between you and the Mark obferved, they are in a ftreight Line, otherwife not. But without all this trouble, they may carry the Chain true enough, if he that follows the Chain always caufe him that goeth before to be in a direct line between himfelf, and the place they are going to, fo as that the Foreman may always cover the Mark from him that goes behind. If they fwerve from the Line, they will make it longer than realy

it

Of Instruments and their Use.

it is, a streight Line being the nearest distance that can be between any two places.

Be sure that they which carry the Chain, mistake not a Chain either over or under in their account, for if they should, the Error would be very considerable; as suppose you was to measure a Field that you knew to be exactly Square, and therefore need measure but one Side of it; if the Chain carriers should mistake but one Chain, and tell you the Side was but 9 Chains when it was really 10, you would make of the Field but 8 Acres and 16 Perches, when it should be 10 Acres just. And if in so small a Line such a great Error may arise, what may be in a greater, you may easily imagine. but the usual way to prevent such Mistakes is, to be provided with 10 small Sticks sharp at one End, to stick into the Ground; and let him that goes before take all into his Hand at setting out, and at the end of every Chain stick down one, which let him that follows take up; when the 10 Sticks are done, besure they have gone 10 Chains; then if the Line be longer, let them change the Sticks, and proceed as before, keeping in Memory how often they change: They may either change at the end of 10 Chains, then the hindmost Man must give the foremost all his Sticks; or which is better, at the end of 11 Chains, and then the last Man must give the first but 9 Sticks keeping one to himself. At every Change count the Sticks, for fear lest you have dropt one, which sometimes happens.

If you find the Chain too long for your use, as for some Lands it is, especially in *America*, you may then take the half of the Chain, and measure as before, remembring still when you put down the

E 4   Lines

Lines in your Field-Book, that you set down but the half of the Chains, and the odd Lines, as if a line measured by the little Chain be 11 Chains 25 Links, you must set down 5 Chain 75 Links, and then in plotting and casting up it will be the same as if you had measured by the whole Chain.

At the end of every 10 Links, you may, if you find it convenient, have a Ring, a piece of Brass, or a Ragg, for your more ready reckoning the odd Links.

When you put down in your Field-Book the length of any Line, you may set it thus, if you please, with a Stop between the Chains and Links, as 15 Chains 15 Links 15.15 or without, as thus, 1515, it will be all one in the casting up.

Of Instruments for the taking of an Angle in the Field.

There are but two material things (towards the the measuring of a piece of Land) to be done in the Field; the one is to measure the Lines (which I have shewed you how to do by the Chain) and the other to take the quantity of an Angle included by these Lines; for which there are almost as many Instruments as there are Surveyors. Such among the rest as have got the greatest esteem in the World, are, the plain Table for small Inclosures, the Semicircle for Champaign Grounds, The Circumferentor, the Theodoite, &c. To describe these to you, their Part, how to put them together, take them asunder, &c. is like teaching the Art of Fencing by Book, one Hours use of them, or but looking on them in the Instrument-maker's Shop, will better describe
them

them to you than the reading one hundred Sheets of Paper concerning them. Let it suffice that the only use of them all is no more (or chiefly at most) but this, *viz.*

To take the Quantity of an Angle

As suppose A B and A C are two Hedges or other Fences of a Field, the Chain serves to measure the length of the Sides A B or A C, and these Instruments we are speaking of are to take the Angle A. And first by the

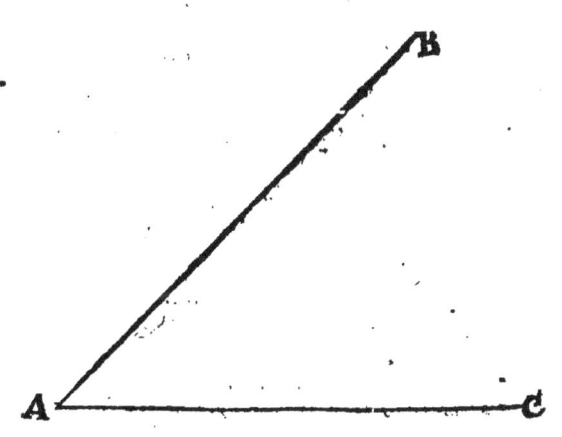

Plain Table.

Place the Table (already fitted for the Work, with a Sheet of Paper upon it) as nigh to the Angle A as you can, the North End of the Needle hanging directly over the *Flower de Luce*; then make a Mark upon the Sheet of Paper at any convenient place for the Angle A, and lay the Edge of the Index to the Mark, turning it about, till through the Sights you espy

espy B, then draw the Line A B by the Edge of the Index. Do the same for the Line A C, keeping the Index still upon the first Mark, then will you have upon your Table an Angle equal to the Angle in the Field.

To take the Quantity of the same Angle by the Semi-circle.

Place your Semi-circle in the Angle A, as near the very Angle as possibly you can, and cause Marks to be set up near B, and C, so far off the Hedges, as your Instrument at A stands; then turn the Instrument about 'till through the fixed Sights you see the Mark at B, there screw it fast; next turn the moveable Index, till through the Sights thereof you see the mark at C, then see what Degrees upon the Limb are cut by the Index; which let be 45, so much is the Angle A B C.

How to take the same Angle by the Circumferentor.

Place your Instrument, as before, at A, with the *Flower de Luce* towards you, direct your Sights to the mark at B, and see what Degrees are then cut by the South end of the Needle, which let be 55; do the same to the mark at C, and let the South end of the Needle there cut 100, substract the lesser out of the greater, the remainder is 45, the Angle required. If the remainder had been more than 180 Degrees, you must then have substracted it out of 360, the last remainder would have been the Angle desired.

This

This laſt Inſtrument depends wholly upon the Needle for taking of Angles, which often proves erroneous; the Needle yearly of itſelf varying from the true North, if there be no Iron Mines in the Earth, or other Accidents to draw it aſide, which in Mountainous Lands are often found: It is therefore the beſt way for the Surveyors, where he poſſibly can, to takes his Angles without the help of the Needle, as is before ſhewed by the Semi-circle: But in all Lands it cannot be done, but we muſt ſometimes make uſe of the Needle, without exceeding great trouble, as in the thick Woods of *Jamaica*, *Carolina*, &c. It is good therefore to have ſuch an Inſtrument, with which an Angle in the Field may be taken either with or without the Needle, as is the Semi-circle, than which I know no better Inſtrument for the Surveyors uſe yet made publick; therefore as I have before ſhewed you, How by the Semi-circle to take an Angle without the help of the Needle; I ſhall here direct you.

How with the Semi-circle to take the Quantity of an Angle in the Field by the Needle.

Screw faſt the Inſtrument, the North End of the Needle hanging directly over the *Flower de Luce* in the Chard, turn the Index about, till through the Sights you eſpy the mark at B; and note what Degrees the Index cuts, which let be 40; move again the Index to the mark at C, and note the Degrees cut, *viz.* 85. Subſtract the Leſs from the Greater, remains 45, the Quantity of the Angle.

Or

Of Instruments and their Use.

Or thus.

Turn the whole Instrument till through the fixed Sights you espy the mark at B, then see what Degrees upon the Chard are cut by the Needle; which for Example are 315, turn also the Instrument till through the same Sights you espy C, and note the Degrees upon the Chard then cut by the Needle, which let be 270; substract the Less from the Greater, (as before in working by the Circumferentor) remains 45 for the Angle. Mark, if you turn the *Flower de Luce* towards the mark, you must look at the North-End of the Needle for your Degrees.

Besides the Division of the Chard of the Semicircle into 360 Equal Parts or Degrees: It is also divided into four Quadrants, each containing 90 Degrees, beginning at the North and South Point, and proceeding both ways till they end in 90 Degrees at the East and West Points; which Points are marked contrary, *viz.* East with a W. and West with an E. because when you turn your Instrument to the Eastward, the end of the Needle will hang upon the West Side, &c.

If by this way of division of the Chard, you would take the aforesaid Angle, direct the Instrument so (the *Flower de Luce* from you) till through the fixed Sights you espy the mark at B; then see what Degrees are cut by the North End of the Needle, which let be N E 44; next direct the Instrument to C, and the North-end of the Needle will cut N E 89; substract the one from the other, and there will remain 45 for the Angle.

But if at the first sight the Needle had hung over N E 55, and at the second S E 80, then take 55 from

from 90, remains 35; take 80 from 90, remains 10, which added to 35, makes 45, the Quantity of the Angle: Moreover, if at the firſt Sight, the North-end of the Needle had pointed to N W 22, and at the ſecond N E 23, theſe two muſt have been added together, and they would have made 45, the Angle as before.

Mark, If you had turned the South part of your Inſtrument to the marks, then you muſt have had reſpect to the South-end of your Needle.

Although I have been ſo long ſhewing you how to take an Angle by the Needle, yet when we come to Survey Land by the Needle, as you ſhall ſee by and by, we need take but half the Pains; for we take not the Quantity of the Angle included by two Lines, but the Quantity of the Angle each Line makes with the Meridian; then drawing Meridian-Lines upon Paper, which repreſent the Needle of the Inſtrument, by the help of a Protractor, which repreſents the Inſtrument, we readily lay down the Lines and Angles in ſuch proportion as their are in the Field.

This way of dividing the Chard into four 90s, is in my Opinion, for any Work the beſt; but there is a greater uſe yet to be made of it, which ſhall hereafter be ſhewed in its proper place.

Of the Field-Book

You muſt always have in readineſs in the Field, a little Book, in which fairly to inſert your Angles and Lines; which Book you may divide by Lines into Columns, as you ſhall think convenient in your Practiſe; leaving always a large Column to the Right-hand, to put down what remarkable things you meet with in your way, as Ponds, Brooks, Mills, Trees, or

the

the like. Thus for Example, if you had taken the Angle A and found it to contain 45 Degrees; and measured the Line AB, and found it to be 12 Chains, 55 Links set it down in your Field-Book thus,

	degrees	Min.	Chain.	Links
A	45	00	12	55

Or if at A you had only turned your fixed Sights to B, and the Needle had then cut 315; in the place of 45, you must have put down 315. If you Survey by Mr. *Norwood*'s way, then there must be four Columns more for E. W. N. and Southing. You may also make two Columns more, if you please, for Off-sets, to the right and left.

Lastly, you may chuse whether you will have any Lines or not, if you can write streight, and in good order, the Figures directly one under another. For this I leave you chiefly to your own fancy; for I believe there are not two Surveyors in *England*, that have exactly the same Method for their Field Notes.

Of the Scale.

Having by the Instrument before spoken of, measured the Angles and Lines in the Field, the next thing to be done, is to lay down the same upon Paper; for which Use the Scale serves. There are several sorts of Scales, some large, some small, according as Men have occasion to use them; but all do principally consist of no more but two sorts of Lines; the first of equal Parts, for the laying down Chains and Links; the second of Chords, for laying down or measuring Angles. I cannot better explain the Scale to you, than by shewing the Figure of such a one as are commonly sold in Shops, and teaching how to use it.   Those

63

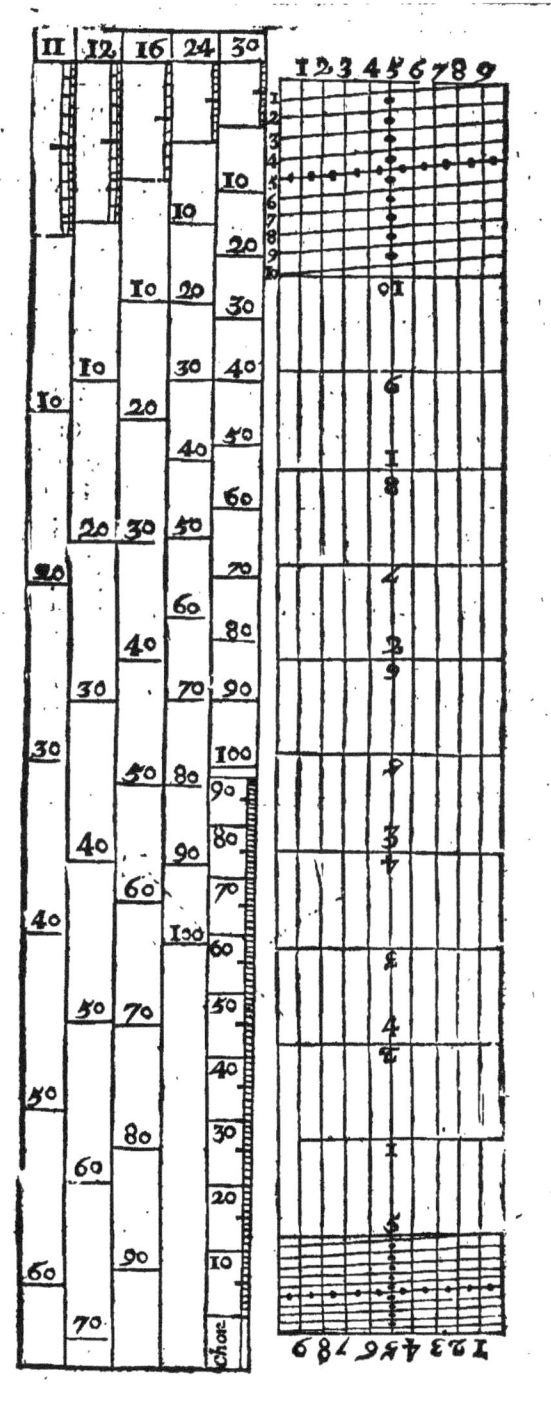

64 Of Instruments and their Use.

Those Lines that are numbred at top with 11, 12, 16, &c. are Lines of equal Parts, containing 11, 12, or 16 Equal parts in an Inch. If now by the Line of 11 in an Inch, you would lay down 10 Chains 50 Links, look down the Line under 11, and setting one foot of your Compasses in 10, close the other till it just touch 50 Links, or half a Chain, in the small Divisions. Then laying your Ruler upon the Paper, by the side thereof make two small Pricks, with the same extent of the Com-
A———B passes, and draw the Line A B, which shall contain in length 10 Chains, 50 Links, by the Scale of 11 in an Inch. The back-side of the Scale, is only a Scale of 10 in and Inch, but divided by Diagonal Lines, more nicely than the other Scales of Equal Parts.

How to lay down an Angle by the Line of Chords

If it were required to make an Angle that should contain 45 Degrees.

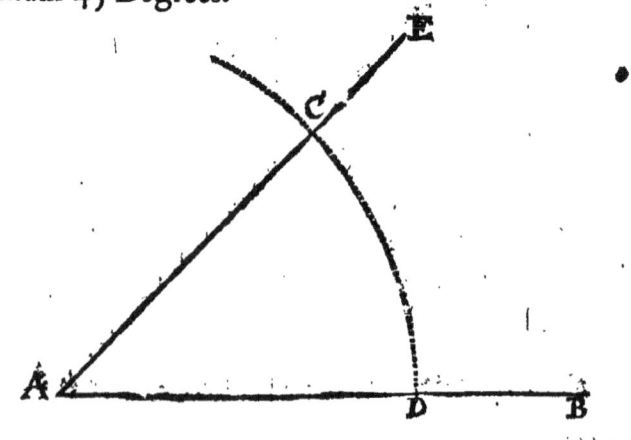

Draw

Draw a Line at pleasure, as A B; then setting one foot of your Compasses at the beginning of the Line of Chords, see that the other fall just upon 60 Degrees: With that extent set one foot in A, and describe the Arch C D. Then take from your Line of Chords 45 Degrees, and setting one foot in D, make a mark upon the Arch at C, through which draw the Line A E: So shall the Angle E A B be 45 Degrees. If by the Line of Chords you would erect a Perpendicular Line, is it no more but to make an Angle that shall contain 90 Degrees.

The reason why I bid you take 60 from the Line of Chords to make your Arch by is, because 60 is the Semi-diameter of a Circle, whose circumference is 360.

How to make a Regular Polygon, or a Figure of 5, 6, 7, 8, or more Sides, by the Line of Chords.

Divide 360, the number of Degrees contained in a Circle, by 5, 6, or 7, the number of Sides you would have your Figure to contain; the Quotient taken from the Line of Choords shall be one Side of such a Figure.

EXAMPLE.

For to make a Pentagon, or a Figure of five Sides: Divide 360 by 5, the Quotient is 72, one side of a Pentagon.

Take 60 Degrees from your Line of Chords, and describe an obscure Circle; which done, take
72 from

72 from your Line of Chords, and beginning at any part of the Circle, set off that extent round the Circle, as from A to B, from B to C, and

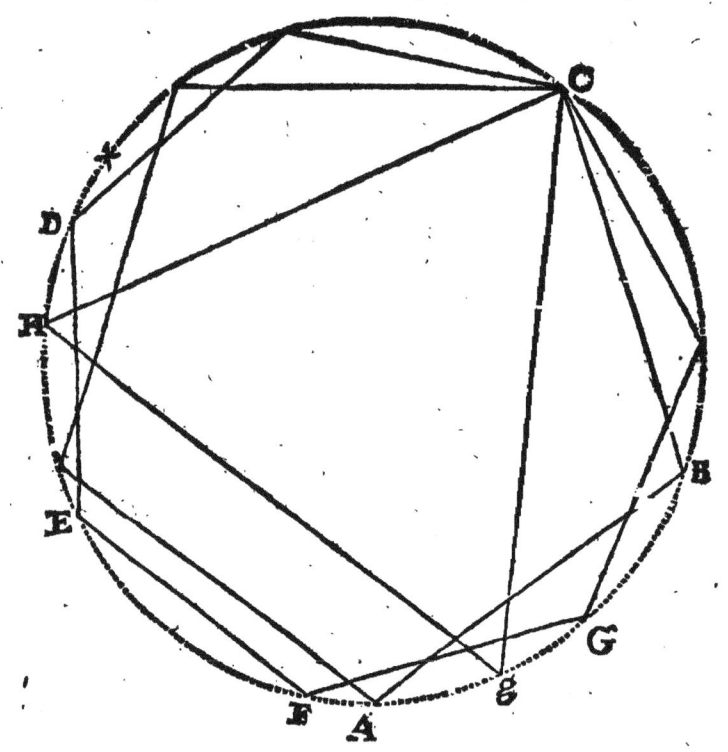

so round till you come to A again. Then having drawn Lines between those Marks, the Pentagon is compleated. The like of any other Polygon, though it contain never so many Sides.

As for Example in a Heptagon: Divide 360 by 7, the Quotient will be 51 Deg. 25 Min. which if you take from the Line of Chords, and set off round the Circle, you will make a Heptagon, as D E, E F, F, G, &c. are the Sides thereof.

To

Of Instruments and their Use. 67

To make a Triangle in a Circle by the Line of Chords.

Firſt take the whole length of your Line of Chords, or the Chord of 90 Degrees with your Compaſſes; which diſtance upon the Circle ſet off from C to *. Then take 30 Degrees from the Line of Chords, and ſet that from * to H. Draw the Line C H, which is one ſide of the greateſt Triangle that can be made in that Circle.

Or you may make it by ſetting off twice the Semi-diameter of the Circle, for 60 and 60 is 120, as well as 90 and 30.

How to make a Line of Chords.

Firſt, make a Quadrant, or the fourth part of a

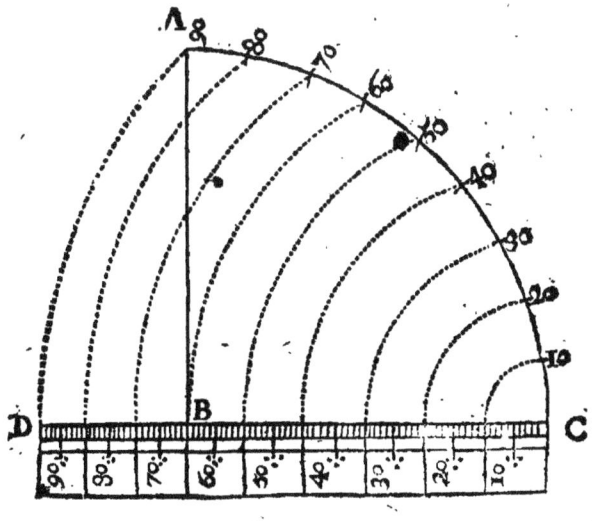

Circle;

Circle, as A, B, C; divide the Arch thereof, viz. A, C, into 90 equal Parts; which you may do by dividing it first into three equal Parts, and every of those Divisions into three Equal Parts more, and every of the last Divisions into ten Equal Parts.

Secondly, Continue the Semi-diameter B C to any convenient length, as to D. Then setting one foot of your Compasses in C, let the other fall on 90. and describe the Arch 90, 90. So likewise 80, 80; 70, 70; and the rest, C D is the Line of Chords, and these Arches cutting it into Unequal Parts, constitute the true Divisions thereof, as you may see by the Figure: You may if you please, draw Lines Parallel to D C, as I have done here, for the better distinguishing every Tenth and Fifth Figure.

Of the Protractor.

The Protractor is an Instrument with which, with more ease and expedition you may lay down an Angle than you can by the Line of Chords: Also when you have Surveyed by the Needle, by placing the Diameter of the Protractor upon a Meridian Line made upon your Paper, you readily with a Needle upon the Arch of the Protractor, prick off the true situation of any Line from the Meridian, without scratching the Paper, as you must do in the use of the Line of Chords. It is made almost like, and graduated together like the Brass Limb of a Semi-circle, performing the same upon Paper, as your Instrument did in the Field: See here the Figure of it.

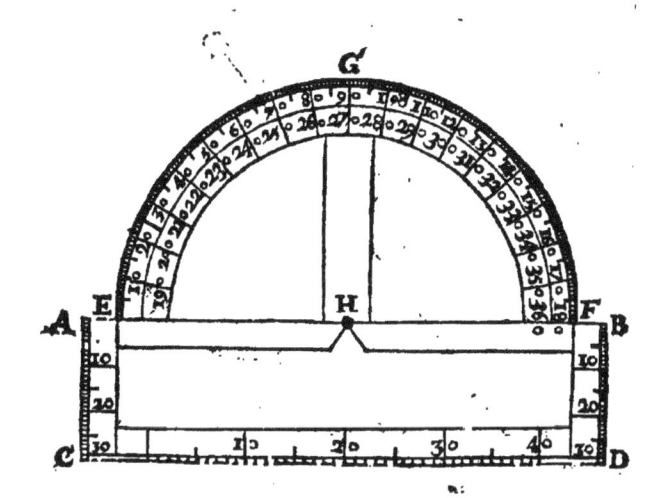

For the use of the Protractor, you must have a fine Needle, such as Women sew withal, put into a small Handle of Wood, or Ivory, or the like, which is to put through the Centre of the Protractor to any Point assigned upon the Paper, that the Protractor may turn round upon it.

How to lay down an Angle with the Protractor.

If it were required by the Protractor to lay down an Angle of 30 Degrees, Draw the Line A B, then take the Protractor, and putting a Needle through the Centre Point thereof, place the Needle in A, so that the Centre of the Protractor may lie just upon the end of the Line at A, move the Protractor about till you find the Diameter thereof lie upon the Line A B; then at 30 Degrees upon the Arch, with your Protracting Needle make a Mark upon the Paper, as at C; draw the Line C A, which shall make an Angle of 30 Degrees, *viz.* B A C.

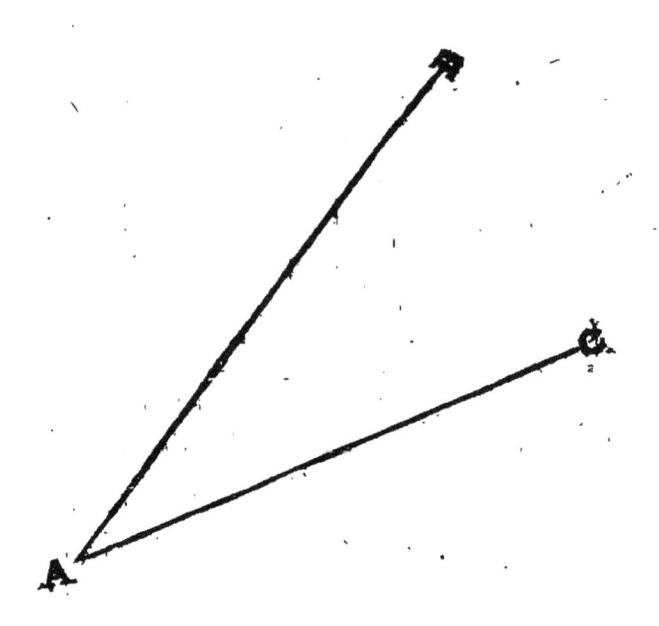

If you Survey according to Mr. *Norwood*'s way before spoken of, it will be good to have the Arch of your Protractor divided accordingly, *viz.* into two Quadrants, or twice 90 Degrees.

I need say no more of a Protractor, any ingenious Man may easily find the several Uses thereof, it being as it were, but only an Epitome of Instruments.

CHAP.

CHAP. VI.

How to take the Plot of a Field at one Station in any Place thereof, from whence you may see all the Angles by the Semi-circle.

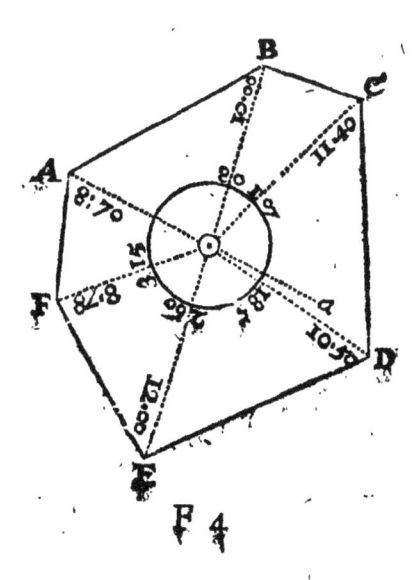

ADmit A B C D E F, to be a Field, of which you are to take the Plot: First set your Semicircle upon the Staff in any convenient Place thereof, as at ☉, and cause Marks to be set up in every Angle: Direct your Instrument, the *Flower de Luce* from you to any one Angle: As for *Example*, to A, and espying the Mark at A through the fixed Sights, there screw fast the Instrument; then turn the movable Index about (the Semi-circle remaining immovable)'till through the Sights thereof you espy the Mark at B. See what Degrees on the Brass Limb are cut by the Index, which let be 80, write that down in your Field-Book, so turn the Index round to every one of the other Angles, putting down in your Field-Book what Degrees the Index points to; as for *Example,* at C 107 Degrees, at D 185; *mark* that at D, the End of the Index will go off the Brass Limb, and the other End will come on; you must therefore look for what Degree, the Index cuts in the innermost Circle of the Limb at E, 260, at F, 315 Degrees.

All which you may note down in your Field-Book thus.

Angles	Degrees	Minutes	Chains . Links
☉ A.	00 .	00 .	8 . 70
☉ B.	080 .	00 .	10 . 00
☉ C.	107 .	00 .	11 . 40
☉ D.	185 .	00 .	10 . 50
☉ E.	260 .	00 .	12 . 00
☉ F.	315 .	00 .	8 . 78

Secondly, Cause the Distance between your Instrument and every angle to be measured, thus from ☉ to A will be found to be 8 Chains 70 Links; from ☉ to B, 10 Chains 00. All which set down in order in your Field-Book, as you see here above; and then have you done what is necessary to be done in that Field towards measuring of it. Your next work is to Protract or lay it down upon Paper.

How

How to Protract the former Observations taken.

First draw a Line at adventure, A *a*, then take from your Scale with your Compasses, the first Distance measured, *viz.* from ⊙ to A 8 Chain 70 Links, and setting one foot in any convenient place of the Line, which may represent the place where the Instrument stood, with the other make a Mark upon the Line as at A, so shall A be the first Angle, and ⊙ the place where the Instrument stood.

Secondly, Take a Protractor, and having laid the Center hereof exactly upon ⊙, and the Diameter or Meridian upon the Line A *a*, the Semi-circle of the Protracture lying upwards. There hold it fast, and with your protracting Pen make a mark upon the Paper against 80 *deg.* 107 *deg. &c.* as you find them in the Field-Book. Then for those Degrees that exceed 180, you must turn the Protractor downward, keeping still the Centre upon ⊙, and placing again the Diameter upon *a* A. Mark out by the Innermost Circle of Divisions the rest of your Observations 185, 260, 315. Then applying a Scale to ⊙, and every one of the Marks, draw the prick'd Lines ⊙ B, ⊙ C, ⊙ D, ⊙ E, ⊙ F.

Thirdly, Take with your Compasses the length of the Line ⊙ B, which you find by the Field-Book to be 10 Chains, which from ⊙ set off to B. The like do for ⊙ C, ⊙ D, and the rest.

Lastly, Draw the Lines A B, B C, C D, *&c.* which will inclose a Figure exactly proportionable to the Field before Surveyed.

How to take the Plot of the same Field at one Station by the Plain Table

Place your Table with a Sheet of Paper upon it at ☉, and making a mark upon the Paper that shall signifie where the Instrument stands, lay your Index to the mark, turning it about till you see through the Sights the mark at A; there holding it fast, draw the Line A ☉. Turn the Index to B, keeping it still upon the first mark at ☉; and when you see through the Sights the mark at B, draw the Line B ☉. Do the same by all the rest of the Angles, and having measured the distance between the Instrument and each Angle, set it off with your Scale and Compasses, from ☉ to A, from ☉ to B, &c. making markes where, upon the several Lines, the distances fall.

Lastly, Between those marks draw Lines, as A B, BC, CD, &c. and then have you the true Plot of the Field ready protracted to your Hand. This Instrument is so plain and easie to be understood, I shall give no more Examples of the Use of it. The greatest Inconveniency that attends it, that when never so little Rain or Dew falls, the Paper will be wet, and the Instrument useless.

How to take the Plot of the same Field at one Station by the Semi-circle, either with the help of the Needle and Limb both together, or by the help of the Needle only.

In the beginning of this Chapter, I shewed you how to take the Plot of a Field at one Station, by the

Divers ways to take the Plots of Fields. 75

the Semi-circle, without respect to the Needle, which is the best way: But that I may not leave you ignorant of any thing belonging to your Instrument, I shall here shew how to perform the same with the help of the Needle two ways. And first with the Needle and Limb together.

Fix the Instrument as before, in ☉, making the North-point of the Needle hang directly over the *Flower de Luce* of the Card; there screw fast the Instrument. Then turn the Index to all the Angles, noting down what Degrees are cut thereby at every Angle, as at A, let be 25, at B, 105, at C, 132; and so of the rest round the Field. And when you have measured the Distances, and are come to Protraction, you must first draw a Line cross your Paper, calling it a North and South Line, which represents the Meridian Line of the Instrument. Then applying the Protractor to that line, mark round the Degrees as they were observed, *viz.* 25, 105, 132, &c. and having set off the Distances, and drawn the outward Lines altogether, like what you were taught at the beginning of the Chapter, you will find the Figure to be the same as there.

Now to perform this by the Needle only, is in a manner the same as the former: For instead of turning the *Index* about the Limb, and seeing what Degrees are cut thereby, here you must turn the whole Instrument about, and observe at every Angle what Degrees upon the Card the Needle hangs over; which set down and Protract as before. But here mind, some Cards are numbred from the North East-wards 10, 20, 30, &c. to 360 *deg.* Some from the North West-ward, which are best for this use, Protractors being made accordingly: For when you

turn

turn your Inſtrument to the Eaſt-ward, the Needle will hang over the Weſt-ward Diviſion, on the contrary.

As for the Uſe of the Diviſion of the Card into four Quadrants, I ſhall ſpeak largely of by and by, therefore for the preſent beg your patience.

How by the Semi-circle to take the Plot of a Field, at one Station, in any Angle thereof, from whence the other Angles may be ſeen.

Let A B C D E F G be the Field, and F the Angle at which you would take your Obſervations. Having placed your Semi-circle at F, turn it about the
North-

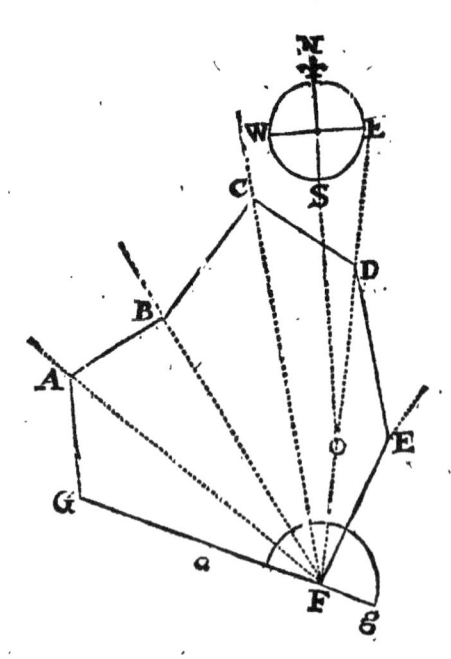

Divers ways to take the Plots of Fields.

North-point of the Card from you, till through the Fixed-Sights, (*Note*, that I call them the fixed-Sights which are on the Fixed-Diameter) you espy the mark at G. Then screw fast the Instrument; which done, move the Index, till through the Sights thereof you see the mark at A, and the Degrees on the Limb there cut by it, will be 20. Move again the Index to the mark at B, where you will find it to cut 40 *deg*. Do the same at C, and it cuts 60 *deg*. Likewise at D 77, and at E 100 *deg*. Note down all these Angles in your Field-Book: Next measure all the Lines, as from F to G 14 Chain, 60 Links; from F to A 18 Chain, 20 Links; from F to B 16 Chain, 80 Links; from F to C 21 Chain, 20 Links; from F to D, 16 Chain, 95 Links; from F to E 8 Chain, 50 Links; and then will your Field-Book stand thus:

Angles	Degrees	Minutes	Chains	Links
G	00	00	14	60
A	20	00	18	20
B	40	00	16	80
C	60	00	21	20
D	77	00	16	95
E	110	00	8	50

To Protract the former Observations.

Draw a Line at adventure, as G *g*, upon any convenient place, on which lay the Centre of your Protractor, as at F, keeping the Diameter thereof right upon the Line G *g*. Then make marks round the Protractor at every Angle, as you find them in the Field-Book, *viz.* against 20, 40, 60, 77, and 100; which

which done, take away the Protractor, and applying the Scale or Ruler to F, and each of the Marks, draw the Lines FA, FB, FC, FD, and FE. Then setting off upon these Lines the true Distances as you find them in the Field-Book; as for the first Line F G 14 Chain, 60 Links, from the second F A 18 Chain, 20 Links, &c. make marks where the end of these Distances fall, which let be at G, A, B, C, &c.

Lastly, Between these Marks, drawing the Lines GA, AB, BC, CD, DE, EF, FG, you will have compleated the Work.

When you Survey thus without the help of the Needle, you must remember before you come out of the Field to make the Meridian-Line, that you may be able to make a Compass shewing the true situation of the Land, in respect of the four Quarters of the Heavens, I mean East, West, North and South; which thus you may do.

The Instrument still standing at F, turn it about till the Needle lies directly over the *Flower de Luce* of the Card; there screw it fast. Then turn the moveable *Index*, till through the Sights you espy any one Angle.

As for Example. Let be D. Note then what Degrees upon the Limb are cut by the Index, which let be 10 *deg*. Mark this down in your Field-Book and when you have protracted as before directed, lay the Centre of your Protractor upon any place of the Line F D, as at ⊙, turning the Protractor about till to 10 *deg*. lie directly upon the Line F D. Then against the end of the Diameter of the Protractor, make a mark as at N, and draw the Line, N ⊙, which is a Meridian, or North and South Line, by which you may make a Compass.

Note,

Divers ways to take the Plots of Fields.

Note, that you may as well take the Plot of a Field at one Station, standing in any Side thereof, as in an Angle: For if you had set your Instrument in *a*, the Work would be the same. I shall forbear therefore (as much as I may) Tautologies.

How to take the Plot of a Field at two Stations, provided from either Station you may see every Angle, and measuring only the Stationary Distance.

Let C D E F G H, be supposed a Field to be measured at two Stations; first when you come into the Field, make choice of two Places for your Stations, which let be as far asunder as the Field will conveniently admit of; also take care that if the Stationary Distance were continued, it would not touch any Angle of the Field; then setting the Semi-circle at A, the first Station, turn it about, the North Point from you, till through the fixed Sights you espy the Mark at your second Station, which admit to be at B, there screw fast the Instrument; then turn the moveable Index, to every several Angle round the whole Field,

and

80 Divers ways to take the Plots of Fields.

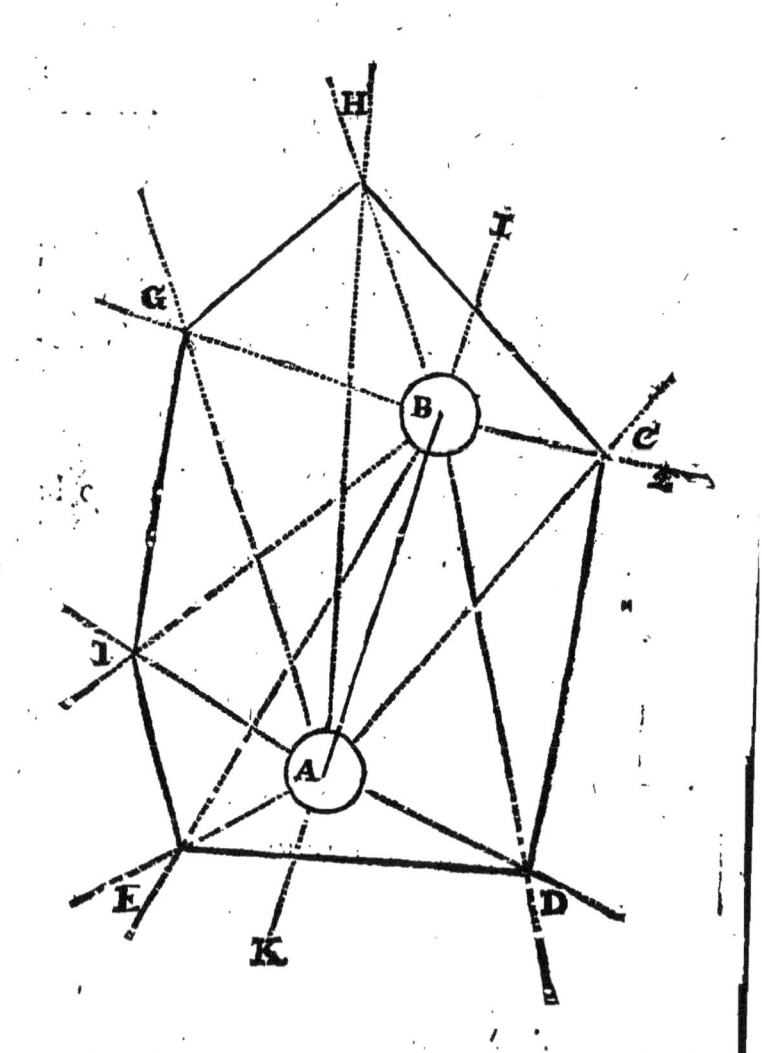

and see what Degrees are cut thereby at every Angle, which note down in your Field-Book as followeth

Angles

Divers ways to take the Plots of Fields. 81

Angles	Degrees	Minutes	
C	24	30	
D	97	00	
E	225	00	First Station.
F	283	30	
G	325	00	
H	346	00	

Secondly, Measure the Distance between the two Stations, which let be 20 Chains, and set it down in the Field-Book.

Stationary Distance 20 *Chains,* 00 *Links.*

Thirdly, Placing the Instrument at B, the Second Station, look backwards through the fixed Sights to the first Station at A, (I mean by looking backward, that the South part of the Instrument be towards A) and having espied the Mark at A, make fast the Instrument, and moving the Index as you did at the First Station to each Angle, see what Degrees are cut by the Index, and note them down as followeth; and then have you done, unless you will take a Meridian Line before you move the Instrument; which you were taught to do a little before.

Angles	Degrees	Minutes	
C	84	00	
D	149	00	
E	194	00	The Second Station.
F	215	00	
G	270	00	
H	322	00	

How to Protract or lay down upon Paper these following Observations.

First, Draw a Line cross your Paper at pleasure, as the Line I K, then take from off the Scale the Stationary distance 20 Chains, and set it upon that Line, as from A to B, so will A represent the First Station, B the Second.

Secondly, apply the Centre of your Protractor, to the Point A, and the Diameter lying streight upon the Line B K; mark out round it the Angles, as you find them in the Field-Book, and through those Marks from A, draw Lines of a convenient length.

Thirdly, Move your Protractor to the Second Station B; and there mark out your Angles, and draw Lines, as before at the first Station.

Lastly, the places where the Lines of the first Station, and the Lines of the Second intersect each other, are the Angles of the Field: As for Example; At the first Station the Angle C was 24 Degrees 30 Minutes; through those Degrees I draw the line A C. At the second Station C was 84 Degrees: Accordingly from the second Station I draw the line B 2; now, I say, where these two lines cut each other, as they do at C, there is one Angle of the Field. So likewise of D E and the rest of the Angles; if therefore between these Intersections you draw streight Lines, as C D, D E, E F, &c. you will have a true Figure of the Field.

This may as well be done by taking two Angles for your Stations, and measuring the line between them;

as

Divers ways to take the Plots of Fields. 83

as C and D, from whence you might as well have seen all the Angles, and consequently as well have performed the Work.

How to take the Plot of a Field at two Stations, when the Field is so irregular, that from one Station you cannot see all the Angles

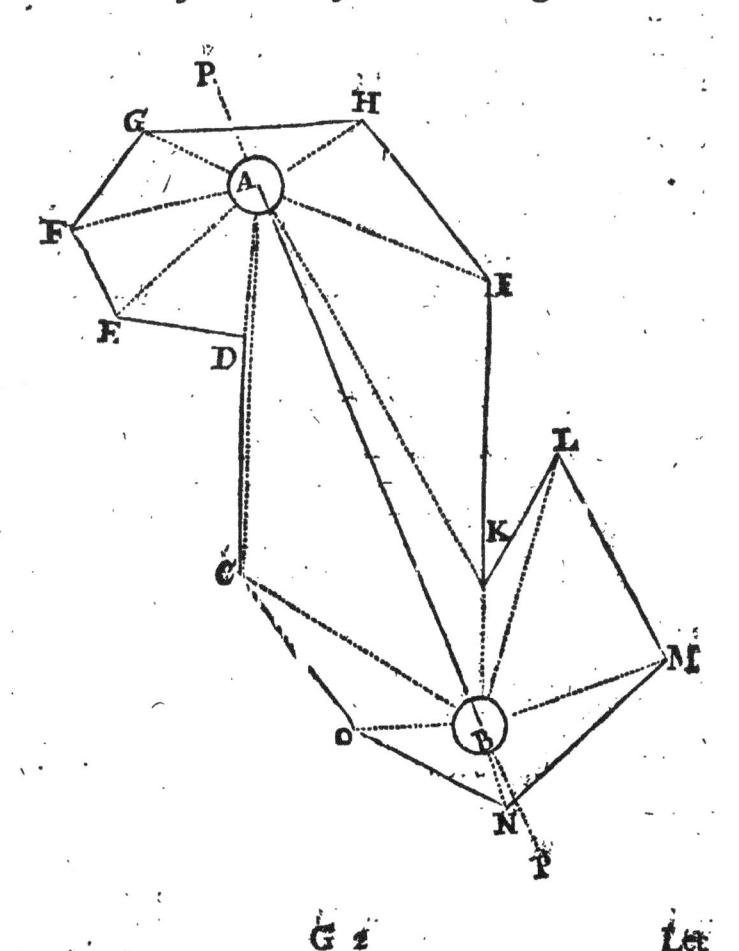

Let CDEFGHIKLMNO be a Field, in which from no one Place thereof all the Angles may be seen; chuse therefore two Places for your Stations, as A and B, and setting the Semi-circle in A, direct the Diameter to the second Station B; there making the Instrument fast, with the Index take all the Angles at that end of the Field, as C D EFGHIK, and measure the Distance between your Instrument and each Angle; measure also the Distance between the two Stations A and B.

Secondly, Remove your Instrument to the second Station at B, and having made it fast so, as that through the Back-Sights you may see the First Station A; take the Angle at that end of the Field, as N O C K L M and measure their Distances also as before, all which done, your Field-Book will stand thus.

First Station.

Angles	Degrees	Minutes	Chains	Links
C	25	00	20	75
D	31	00	8	10
E	67	00	9	85
F	101	00	10	80
G	137	00	7	00
H	263	00	6	70
I	316	00	13	70
K	354	00	24	50

The Distance between the two Stations 31 Chains, 60 Links.

Second

Second Station.

Angles	Deg.	Min.	Chain.	Link.
N	3	30	4	20
O	111	00	7	00
C	145	00	15	60
K	205	00	7	48
L	220	00	15	00
M	274	00	11	20

To lay this down upon Paper, draw at adventure the Line PBAP; then taking in with the Compasses the distance between the two Stations, *viz.* 31 Chains, 60 Links; set it upon the Line, making Marks with the Compasses as A and B, A being the First Station, B the Second, lay the Protractor to A, the North-end of the Diameter towards B, and mark out the several Angles observed at your first Station, drawing Lines, and setting off the Distances as you were taught in the beginning of this Chapter, *Fig.* I.

Do the same at B, the second Station; and when you have marked out all the Distances, between those Marks, draw the Bound-lines.

I am the briefer in this, because it is the same as was taught concerning *Fig.* I. for if you conceive a Line to be drawn from C to K; then would there be two distinct Fields to be measured, at one Station a-piece.

If a Field be very irregular, you may after the same manner make three, four or five Stations, if you please; but I think it better to go round such a Field and measure the bounding Lines thereof: Which by and by I shall shew you how to do.

Note,

86 Divers ways to take the Plots of Fields.

Note, in the foregoing Figure you might as well have had your Stations in two convenient Angles, as D and K, and have wrought as you were taught concerning *Fig. 2*, the Work would have been the same.

How to take the Plot of a Field at one Station in an Angle (so that from that Angle you may see all the other Angles) by measuring round about the said Field.

ABCDE is the Field, and A the Angle appointed for the Station; place your Semi-circle in A, and direct the Diameter thereof till through the fixed Sights you see the Mark at B, then screw it fast, and turn the Index to C, observing what Degrees are there cut upon the Limb; which let be 68 Degrees; turn it further, till you espy D, and note

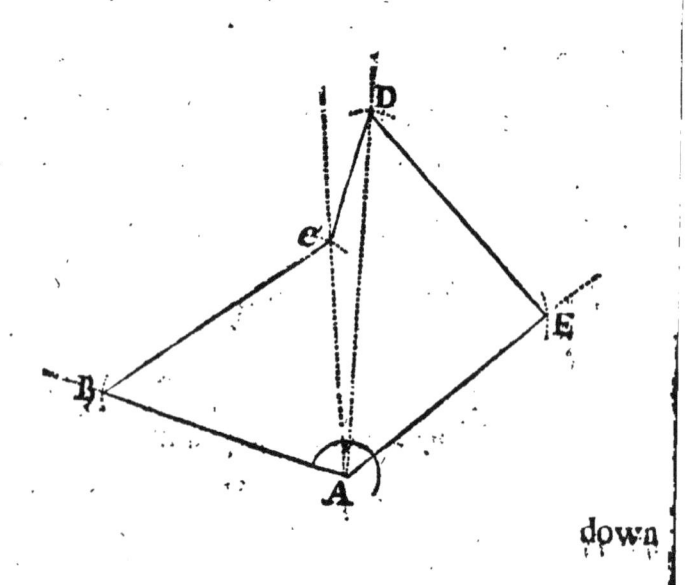

down

Divers ways to take the Plots of Fields.

down the Degrees there cut, *viz.* 76 Degrees; do the like at E, and the Index will cut 124 Degrees: This done, measure round the Field, noting down the length of the Side-lines between Angle and Angle, as for A to B, 14 Chains oo Links, from B to C, 15 Chains oo Links; from C to D, 7 Chains oo Links; from D to E, 14 Chains 40 Links; and from E to A, 14 Chains 05 Links.

Then will your Field-Book be as hereunder.

Angles	Degrees	Minutes		Chains	Links
C—	68	00	A B—	14	00
D—	76	00	B C—	15	00
E—	124	00	C D—	07	00
			D E—	14	40
			E A—	14	05

To protract which, draw the Line A B at adventure, and applying the Centre of the Protractor to A, (the Diameter lying upon the Line A B, and the Semicircle of it upwards) prick off the Angle, as against 68 : 76 : and 124 : make Marks, through which Marks draw the Lines A C, A D, A E, long enough befure; then take in with your Compasses, from off the Scale, the length of the Line AB, *viz.* 14 Chains, and setting one foot of the Compasses in A, with the other cross the line, as at B; also for B C take in 15 Chains, and setting one Foot in B, with the other cross the line A C, which will fall to be at C; for the line C D take in 7 Chains, and setting one Foot in C, cross the Line A D, *viz.* at D; then for DE, take in 14 Chains 40 Links, and set-

ting

ting one foot of the Compasses in D E, with the other cross the Line A E, which will fall at E: Lastly for EA take 14 Chains 5 Links with your Compasses, and setting one Point in E, see if the other fall exactly upon A, if it does, you have done the Work true, if not, you have erred; between the Crosses or Intersections, draw streight Lines, which shall be the bounds of the Field, *viz.* A B, B C, C D, D E, E A.

How to take the Plot of the foregoing Field, by measuring one Line only, and taking Observations at every Angle.

Being as you have been just before taught, till you have taken the Angles, CDE, *viz.* 68, 76, and 124 Degrees; then leaving a good Mark at A, which may be seen all round the Field, go to B, measuring as you go the Distances from A to B, which is all the Lines you need to measure; and planting your Semicircle at B, direct the South part thereof toward A, until through the back fixed Sights you see the Mark at A, there making it fast, turn the Index about till you espy C, and note down the Degrees there cut, which let be 129 Degrees; move your Instrument to C, and still keeping the South part of the Diameter to A, turn the Index to D, where it will cut 20 Degrees; then remove to D, and espying A through the Back-Sights, turn the Index to E, where it will cut 135 Degrees. Note all this in your Field-Book.

Angles

Divers ways to take the Plots of Fields. 89

{ Angles taken at the First Station }   { Angles round the Field. }
C — 68 ⎫   B . 129 ⎫
D — 76 ⎬ Degrees   C : 20 ⎬ Degrees
E — 124 ⎭   D . 135 ⎭
   Line A B : 14 Chains.

To protract this you must work as you were taught concerning the foregoing Figure, until you have drawn the Lines AB, AC, AD, AE, and set off the Line AB, 14 Chains; then laying the Centre of your Protractor to B, and the South-end of the Diameter (or that marked with 180 Degrees) towards A, make a Mark against 129 Degrees, and through that mark from B, draw the Line BC, till it intersect the Line AC, which it will do at C: Lay also the Centre of the Protractor upon C, the Diameter thereof upon A C, and against 20 Degrees make a Mark, through which from C, draw the line C D, till it intersect the Line A D, which it will do at D; lastly place your Protractor at D, the Diameter thereof upon the Line DA, and make a mark against 135 Degrees, through which mark draw the Line D E, until it intersect the Line AE at E, also drawing the Line E A, you have done.

This may be done otherwise thus, after you have, standing at A, taken the several Angles, and measured the Distance AB, you may only take the quantity of the bounding Angles, without respect to A; As the Angle at B is 51 Degrees, at C, (an outward Angle, which in your Field-Book you should distinguish with a Mark ⋗) 138; and so of the rest. And when you come to Plot, having found the
place

90 Divers ways to take the Plots of Fields.

place for B, there make an Angle of 51 Degrees, drawing the Line till it interfect A C, &c.

You may alfo Survey a Field after this manner, by fetting up a mark in the middle thereof, and meafuring from that to any one Angle, alfo in the Obfervations round the Field, having refpect to that Mark, as you had here to the Angle A.

It is too tedious to give Examples of all the Varieties; befides, it would rather puzzle than inftruct a Neophyte.

How to take the Plot of a Large Field or Wood, by meafuring round the fame, and taking Obfervations at every Angle thereof by the Semi-circle.

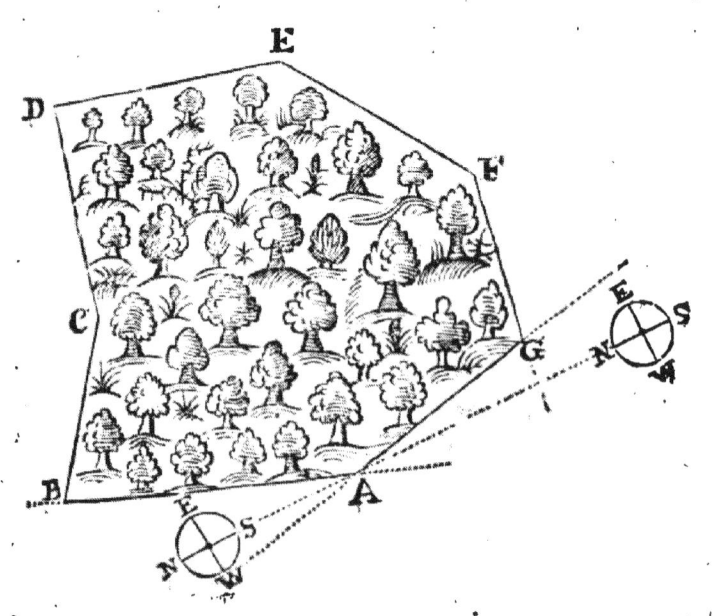

Suppofe

Divers ways to take the Plots of Fields.

Suppose A B C D E F G, to be a Wood, through which you cannot see to take the Angles, as before directed, but must be forced to go round the same; first plant the Semi-circle at A, and turn the North-end of the Diameter about, till through the fixed Sights you see the Mark at B, then move round the Index, till through the Sights thereof you espy G, the Index there Cutting upon the Limb 146 Degrees.

2. Remove to B; and as you go, measure the Distance A B, *viz.* 23 Chains 40 Links; and planting the Instrument at B, direct the North-end of the Diameter to C, and turn the Index round to A, it then pointing then to 76 Degrees.

3. Remove to C, measuring the Line as you go, and setting your Instrument at C, direct the North-end of the fixed Diameter to D, and turn the Index till you espy B, and the Index then cutting 205 Degrees; which, because it is an outward Angle, you may mark thus > in your Field-Book.

4. Remove to D, and measure as you go; then placing the Instrument at D, turn the North-end of the Diameter to E, and the Index to C, the Quantity of the Angle will be 84 Degrees.

And thus you must do at every Angle round the Field as at E, you will find the Quantity of that Angle to be 142 Degrees; F 137; G 110: But there is no need for your taking the last Angle, nor yet measuring the two last Sides, unless it be to prove the Truth of your Work; which is indeed convenient. When you have thus gone round the Field, you will find your Field-Book to be as followeth,

Angles

92 *Divers ways to take the Plots of Fields*

Angles		Lines	
Deg.	Min.	Ch.	Link.
A . 146	. 00	A B . 23	. 40
B . 76	. 00	B C . 15	. 20
C . 205	. 00	C D . 17	. 90
D . 84	. 00	D E . 20	. 60
E . 142	. 00	E F . 18	. 85
F . 137	. 00	F G . 13	. 60
G . 110	. 00	G A . 19	. 28

To protract this, draw a dark Line at adventure, as AB; upon which set off the Distance, as you see in your Field-Book, 23 Chains 40 Links, from A to B; then laying the Centre of your Protractor upon A, and the Diameter upon the Line AB, the North-end, or that of oo Degrees towards B; on the out-side of the Limb make a Mark against 146 Degrees, through which Mark from A draw the Line A G; so have you the first Angle and first Distance.

2. Place the Centre of the Protractor upon B, and turn it about until 76 Degrees lies upon the line AB; there hold it fast, and against the North-end of the Diameter make a Mark, through which draw a line, and set off the Distance BC 15 Chains 20 Links.

3. Apply the Centre of the Protractor to C, (the Semi-circle thereof outward, because you see by the Field-Book it is an outward Angle) and turn it about till 205 Degrees, lie upon the Line CB; then against the Upper or South-end of the Diameter make a Mark, through which draw a Line, and set off 17 Chains 90 Links from C to D. 4. Put

Divers ways to take the Plots of Fields.

4. Put the Centre of the Protractor to D, and make 84 Degrees thereof lie upon the line CD; then making a mark at the end of the Diameter or o *deg.* Through that mark draw a line, and set off 20 Chains, 60 Links, *viz.* D E.

5. Move the Protractor to E, and make 142 *deg.* to lie upon the line E D. Then at the end of the Protractor, make a mark as before, and setting off the distance 18 Chains 85 Links, draw the line E F.

6. Lay the Centre of the Protractor upon F, and making 137 *deg.* lie upon the line EF; against the end of the Diameter make a mark, through which draw the line F G, which will intersect the line A G at G: So have you a true Copy of the Field or Wood: But you may, if you think fit to prove your Work, set off the distance from F to G; and at G apply your Protractor, making 110 *deg.* thereof to lie upon the line FG. Then if the end of the Diameter point directly to A, and the distance be 90 Chain, 28 Links, you may be sure you have done your Work true.

Whereas I bid you put the North-end of the Instrument and of the Protractor towards B, it was chiefly to shew you the variety of Work by one Instrument; for in the Figure before this, I directed you to do it the contrary way; and in this Figure, if you had turned the South-end of the Instrument to G, and with the Index had taken B, and so of rest, the Work would have been the same, remembring still to use the Protractor the same way as you did your Instrument in the Field.

Also, if you had been to have Surveyed this Field or Wood by the help of the Needle; after you had planted the Semicircle at A, and posited it, so

that

that the Needle might hang directly over the *Flower de Luce* in the Card, you should have turned the Index to B, and put down in your Field-Book what Degrees upon the Brass Limb had then been cut thereby; which let be 20. Then moving your Instrument to B, make the Needle hang over the *Flower de-Luce*, and turn the Index to C, and note down what *degrees* are there cut. So do by all the rest of the Angles. And when you come to Protract, you must draw lines Parallel to one another cross the Paper, not farther distant than the breadth of the Parallelogram of your Protractor; which shall be Meridian Lines, marking one of them at one end N, for the North, and at the other S, for South. This done, chuse any place which you shall think most convenient upon one of the Meridian lines for your first Angle at A; and laying the Diameter of your Protractor upon that Line, against 20 *deg.* make a mark; through which draw a line, and upon it set off the distance from A to B.

In like manner proceed with the other Angles and Lines, at every Angle laying your Protractor Parallel to a North and South Line; which you may do by the Figures graduated thereon, at either end alike.

When you have Surveyed after this manner, how to know before you go out of the Field whether you have wrought true or not.

Add the Sum of your Angles together, as in the Example of the precedent Wood, and they make 900. Multiply 180 by a number less by two than

Divers ways to take the Plots of Fields.

the number of Angles; and if the Product be equal to the Sum of the quantity of all the Angles then have you wrought true. There were seven Angles in that Wood, therefore multiply 180 by 5, and the Product is 900.

If you Survey, by taking the quantity of every Angle, and if all be inward Angles, you must work as before. But if one or more be outward Angles, you must substract them out of 180 *deg.* and add the Remainder only to the rest of the Angles. And when you multiply 180 by a Sum less by 2 than the number of your Angles, you are not to account the outward Angles into the number. Thus in the precedent Example I find one outward Angle, *viz.* C 205; the quantity of which, if it had been taken, would have been but 155 *deg.* That taken from 180 *deg.* there remains 25; which I add to the other Angles, and they make then in all 720. Now because C was an outward Angle, I take no notice of it, but see how many other Angles I have, and I find 6; a number less by two than 6, is 4; by which I multiply 180, and the Product is 720, as before.

Directions to measure Parallel to a Hedge (when you cannot go in the Hedge itself.) and also in such case, how to take your Angles.

It is impossible for you when you have a Hedge to measure, to go at top of the Hedge itself; but if you go Parallel thereto, either within or without, and make your Parallel-line of the same length

as

as the Line of your Hedge, your Work will be the same. Thus if A B was a bushy Hedge, to which

You could not conveniently come nigher to plant your Instrument than ⊙; let him that goes to set up your mark at B, take before he goes the distance, A ⊙, which he may do readily with a Wand or Rod; and at B let him set off the same distance again, as to ✚, where let the mark be placed for your Observation; and when the Chain bears, measure the distance ⊙ ✚, be sure they have respect to the Hedge AB, so as that they make ⊙ ✚ equal to AB, or of the same length.

But to make this more plain, suppose ABC to be a Field; and for the Bushes, you cannot come nigher than ⊙ to plant your Instrument. Let him that sets

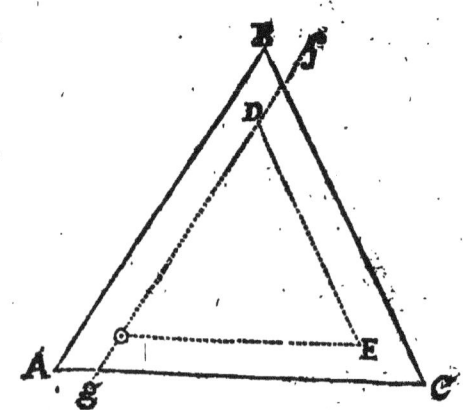

up the Marks, take the distance between the Instrument ⊙ and the Hedge AB; which distance let him set of again nigh B, and set up his Mark at D; likewise

Divers ways to take the Plots of Fields 97

wise let him take the distance between ☉ and the Hedge A C, and accordingly set up his Mark at E. Then taking the Angle *d* ☉ E, it will be the same as the Angle B A C: So do for the rest of the Angles. But when the Lines are measured, they must be measured of the same length as the outside Lines, as the Line ☉ *d*, measured from g to f, *&c*. The best way therefore is for them that measure the Lines, to go round the Field on the outside thereof, although the Angles be taken within.

How to take the Plot of a Field or Wood, by observing near every Angle, and measuring the Distance between the Marks of Observation, by taking, in every Line, two Off-sets to the Hedge.

Let A, B, C, D, be a Wood or Field, to be thus measured. Cause your Assistants to set up Marks in

every Angle thereof, not regarding the distance from the Hedges, so much as the convenience for planting
H the

to be. To begin therefore, plant your Semi-circle in any convenient place of the Field, for taking a large Square as at 1; and laying the moveable Index upon 90 *deg.* look through the Sights, and cause a Mark to be set up in that Line, as at 4 : Looking also through the Fixed-Sights, cause another Mark to be set up as at 2. Measure out from your Instruments towards either of these Marks, any number of Chains, as from 1, to 2, 12 Chains; from 1, 4, 12 Chains. But as you measure, remember to take the Off-sets in a Perpendicular-Line to every Angle or Side, if there be occasion, as here 7, which is 1 Chain, 50 Links; from my Stations I take an Off-set to a side of the Hedge and put it down accordingly 5 Chains, 40 Links. So at 8 I take an Off-set to an Angle, *viz.* 8 B, 6 Chains, which Off-set is at the end of 8 Chains, 30 Links in my first Line. Then seeing in that Line there is no more occasion of Off-sets, I measure on to 2, making the Line 1, 2, 12 Chains. Then planting my Instrument at 2, I direct the fixed-Sights to my first Station, and laying the Index upon 90 *deg.* I cause a Mark to be set up, so as that I may see it through the Sights; and upon that Line, as I measure out 12 Chains, I take the Off-sets C 9, D 10. In like manner you must do for the other Angle, Lines and Off-sets.

And when you have thus laid out your Square and taken all your Off-sets, you will find in your Field-Book such *Memorandums* as these, to help you to Protract.

Divers ways to take the Plots of Fields. 101

The Angles 4 *Right-Angles.*
The Sides 12 *Chains,* 00 *Links each.*

I went round *cum Sole*, or the Hedges being on my Left-hand.

	C.	L.		C.	L.
In the first Line, at	1	50	Off-set to a Side-Line	5	40
	8	30	Off-set to an Angle	6	00

	C.	L.		C.	L.
In the second Line, at	3	50	Off-set to an Angle	6	00
	10	70	Off-set to an Angle	5	50

	C.	L.		C.	L.
In the third Line, at	10	00	Off-set to an Angle	5	30

	C.	L.		C.	L.
In the fourth Line, at	4	30	Off-set to an Angle	4	40
	6	70	Off-set to an Angle	1	50
	10	80	Off-set to a Side	2	20

Now to lay down upon Paper the foregoing Work, make first a Square Figure, whose Side may be 12 Chains, as 1, 2, 3, 4. Then considering you went with the Sun, take 1, 2, for the first Line; and taking from your Scale 1 Chain, 50 Links, set it upon the Line from 1 to 7: at 7 raise a Perpendicular, as 7, 6, making it according to your Field-Book, 5 Chains, 40 Links long. Also for the second Off-set upon the

H 3 same

102 Divers ways to take the Plots of Fields.

fame Line, take from your Scale of Equal Parts 8 Chains, 30 Links, which set upon the Line from 1 to 8, and upon 8 make the Perpendicular-line 8 B, 6 Chains in length.

For the Off-sets of the second Line, take 3 Chains, 50 Links from the Scale, and set it from 2 to 9; at 9 make a Perpendicular-Line 6 Chains long, *viz.* 9 C: Also for the second Off-set of the same Line, take 10 Chains 70 Links, and set it from 2 to 10; at 10 make the Perpendicular 10 D, 5 Chains, 50 Links in Length.

For the Off-sets of the third Line, take from your Scale 10 Chains, and set it up from 3 to 11; and at 11 make the Perpendicular 11 E, 5 Chains, 30 Links long.

For the Off-sets of the fourth Line, take from your Scale 4 Chains, 30 Links, and set it from 4 to 12; and at 12 make the Perpendicular 12 F, 4 Chains, 40 Links long. Also take 6 Chains, 70 Links and set it from 4 to 13; and at 13 make the Perpendicular 13 G, 1 Chain, 50 Links long.

Lastly, take 10 Chains, 80 Links, and set it from 4 to 1, and at I, make the Perpendicular 1, 5, 2 Chains, 20 Links long.

Then have you no more to do, but through the ends of these Perpendiculars to draw the Bounding-lines, remembring to make Angles where the Field-Book mentions Angles; and were it mentions Side-lines, there to continue such Side-lines till they meet in an Angle.

Although I mention a Square, yet you are not bound to that Figure; for you may with the same success use a Parallelogram, Triangle, or any other Figure. Nor are you bound to take the Off-sets in

Perpen-

Divers ways to take the Plots of Fields.

perpendicular-lines, although it be the best way; for you may take the Angles with the Index, from any part of the Line.

This way was chiefly intended for such as were not provided with Instruments; for instead of the Semi-circle with a plain Cross only, you may lay out a Square, the rest of the Work being done with a Chain.

How by the help of the Needle to take the Plot of a large Wood by going round the same, and making use of that Division of the Card that is numbred with four 90's or Quadrants.

Let A B C D E represent a Wood; set your Instrument at A, and turn it about till through the Fixed Sights you espy B, then see what Degrees in the Division before spoken of, the Needle cuts, which let be N 7 W, measure A B 27 Chains, 70 Links; then setting the Instrument at B, direct the Sights to C, and see what then the Needle cuts, which let be N. 74 E; measure B C 39 Chains, 50 Links in like manner measure every Line, and take every Angle, and then your Field-Book will stand thus; as followeth hereunder.

H 4   Lines

104 Divers ways to take the Plots of Fields.

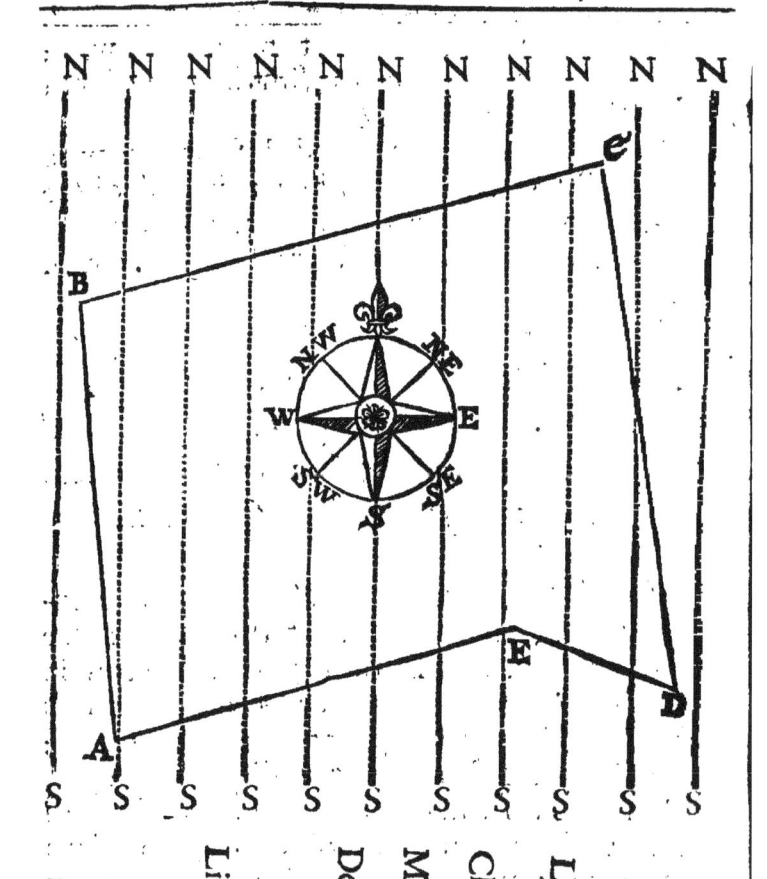

Lines · Degrees · Minutes · Chains · Links

AB : N.W : 7 : 00 : 28 : 20
BC : N. E : 74 : 00 : 39 : 50
CD : S. E : 9 : 00 : 38 : 00
DE : N.W : 63 : 20 : 14 : 55
EA : S. W : 74 : 80 : 28 : 60

To

Divers ways to take the Plots of Fields. 105

To lay down which upon Paper, draw Parallel Lines through your Paper, which shall reprefent Meridians or North and South Lines, as the Lines N S N S; then applying the Protractor (which should be graduated accordingly, with twice 90 Degrees, beginning at each end of the Diameter, and meeting in the middle of the Arch) to any convenient place of one of the Lines as to A, lay the Meridian line of the Protractor to the Meridian-line on the Paper, and againſt 7 *deg.* make a Mark, through which draw a line, and ſet off thereon the Diſtance A B 28 Chains 20 Links. Secondly, apply the Centre of the Protractor to B, and (turning the Semi-circle thereof the other way, becauſe you ſee the Courſe tends to the Eaſtward) make the Diameter thereof lie parallel to the Meridian lines on the Paper, (which you may do by the Figures at the ends of the Parallelogram) and againſt 74 Degrees make a Mark, and ſet off 39 Chains 50 Links, and draw the line B C; the like do by the other Lines and Angles, until you come round to the place where you began

This is the moſt uſual way of plotting Obſervations taken after this manner, and uſed by moſt Surveyors in *America*, where they lay out very large Tracts of Land: But there is another way, though more tedious, yet ſurer; (I think firſt made publick by Mr. *Norwood*) whereby you may know before you come out of the Field, whether you have taken your Angles, and meaſured the Lines truly or not, and is as followeth.

When you have Surveyed the Ground as above directed, and find your Field-Book to ſtand as before; caſt up what Northing, Sonthing, Eaſting or Weſt-
ing

106 Divers ways to take the Plots of Feilds.

ing every line makes; that is to say, How far at the End of every line you have altered your Meridian, and what distance upon a Meridian-line you have made: As for Example, Suppose A B was the Side of a Field measured to be 20 Chains, N S a Meridian-

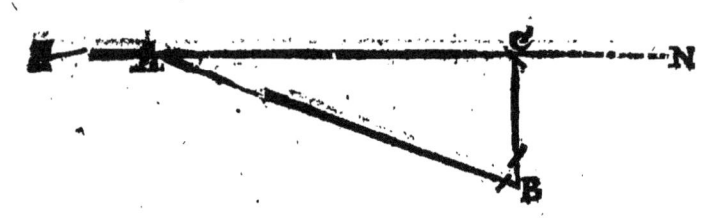

line, the Angle C A B North 20 *deg* East. The business is to find the length of the line A C, which is called the Northing, or the difference of Latitude; also the length of the line C B, which is called the Easting, or difference of Longitude; which you may do indifferently true by laying them down thus upon Paper. But passing this and the *Gunter*'s Scale, the only way is by the Tables of Sines and Logarithms, where the Proportion is this.

As Radius or Sine of 90 Degrees, *viz.* the Right Angle C is to the Logarithm of the line A B 20 Chains;

So is the Sine of the Angle C A B 20 Degrees to the difference of Longitude C K 6 Chains 80 Links.

Secondly, To find the difference of Latitudes, or the Line A C, say,

As Radius is to the Logarithm of the line A B 20 Chain, so is the Sine Complement of the Angle at A to the Logarithm of the Line A C 18 Chains 80 Links:

Example

Example of the foregoing Figure.

In the precedent Figure, I find in my Field-Book the first Line to run N. W. 7 Degrees 28 Chain 20 Links; now to find what Northing, and what Westing is here made, I say thus,

As Radius   10,000000
Is to the Logarithm of the Line 28 Chains 20 Links,   1,450249
So is the Sine of the Angle from the Meridian, viz. 7 Degrees   9,085894
To the Logarithm of the Westing 3 Chains 43 Links   10,536143

Again,

As Radius   10,000000
Is to the Logarithm 28 Chains 20 Links   1,450249
So is the Sine Complement of 7 Degrees   9,996751

To the Log. of the Northing 27 Ch. 99 Lin. 1,1447000

And having thus found the Northing and West- of that Line: I put it down in the Field-Book against the Line under the proper Titles N. W. in like manner I find the Latitude and Longitude of all the rest, and having set them down, the Field-Book will appear thus.

Divers ways to take the Plots of Fields.

Lines		Degrees	Chains Minutes	Links	N	S	E	W
AB	NW	7 : 00	28 : 20	27 : 99	.. : ..	.. : ..	03 : 43
BC	N E	74 : 00	39 : 50	10 : 89	.. : ..	.. : ..	37 : 97	.. : ..
CD	S E	9 : 00	38 : 00	.. : ..	37 : 53	05 : 95	.. : ..
DE	NW	63 : 20	14 : 55	06 : 53	.. : ..	.. : ..	13 : 00
EA	S W	74 : 00	28 : 60	.. : ..	07 : 88	.. : ..	27 : 49
					45 : 41	45 : 41	43 : 92	43 : 92

This done add the Northings together, also all the Southings, and see if they agree; also all the Eastings and Westings; and if they agree likewise, then you may be sure you have wrought truly, otherwise not. Thus in this Example the sum of the Northings is 45 Chains 41 Links; so likewise is the summ of the Southings; also the summ of the Eastings, is 43 Chains 92 Links, so is the summ of the Westings: therefore I say I have surveyed that Piece of Land true.

But because this way of casting up the Northing, Southing, Easting or Westing of every Line may seem tedious and troublesome to you; I have at the End of this Book, made a Table, wherein by Inspection only, you may find the Longitude and Latitude of every line, what quantity of Degrees soever it is situated from the Meridian.

Another way of plotting the foregoing piece of Ground according to the Table in the Field-Book of N S, E W, is as followeth.

Draw

Divers ways to take the Plots of Feilds.

Draw a line at adventure; as the line n ⊙ A S for a Meridian-line; then beginning in any place of that line, as at A, set off the Northing of the first line, as from A, to ⊙ 1, *viz.* 27 Chain 99 Links; then taking with your Compasses the Westings of the same line, *viz.* 3 Chains 43 Links; set one Foot in ⊙ 1, and with the other make the Arch *a a*; next take the length of your first Line, as you find it in the Field-Book, *viz.* 28 Chains 20 Links; and setting one Foot of the Compasses in A, with the other

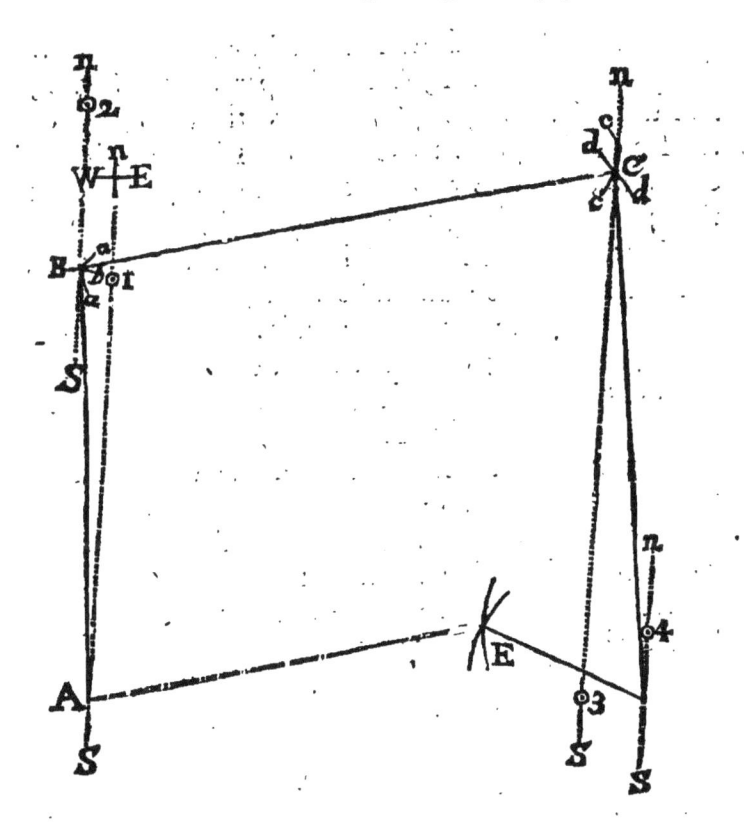

cross

the Inſtrument, ſo as you may ſee from one Mark to another. Then beginning at ⊙ 1, take the quantity of that Angle, and meaſure the diſtance 1, 2. But before you begin to meaſure the Line, take the Off-ſet to the Hedge, *viz*. the diſtance ⊙ *e*; and in taking of it, you muſt make that little Line ⊙*e* perpendicular to 1. 2; which is eaſy done, when your Inſtrument ſtands with the Fixed Sights towards 2, by turning the moveable Index till it lie upon 90 *deg*. which then will direct to what place of the Hedge to meaſure, as *e*, that little Line ⊙ *e*, ſet down in your Field-Book under Title *Off-ſet*. So likewiſe when you come to 2, meaſure there the Off-ſet, again, *viz*. ⊙ *f*. Then taking the Angle at 2, meaſure the Line 2, 3, and the Off-ſets 2 *g*, 3 *h*. The like do by all the reſt of the Lines and Angles in the Field, how many ſoever they be. And when you come to lay this down upon Paper; firſt, as you have been taught before, Protract the Figures 1, 2, 3, 4. That done, ſet off your Off-ſets as you find them in the Field-Book, *viz*. ⊙ *e*, and ⊙ *f*, perpendicular to the Line 1, 2; alſo ⊙ *g*, ⊙ *h*, perpendicular to the Line 2, 3, making Marks at *e*, *f*, *g*, *h*, and the reſt; through which draw Lines that ſhall interſect each other at the true Angles, and deſcribe the true Bound-Lines of the Field or Wood.

In working after this manner, obſerve theſe two Things. Firſt, if the Wood be ſo thick, that you cannot go on the inſide thereof, you may after the ſame manner as well perform the Work, by going on the out-ſide round the Wood.

Secondly, if the Lines are ſo long, that you cannot ſee from Angle to Angle, cauſe your Aſſiſtant to ſet up a Mark ſo far from you as you can conveniently

Divers ways to take the Plots of Fields.

ently see it, as at n. Measure the distance ☉ 1 n, and take the Off-set from n to the Hedge. Then at n turn the Fixed-sights of the Instrument to ☉ 1, and by that Direction, proceed on the Line till you come to an Angle.

This way of Surveying is much easier done (though 1 cannot say truer) by taking only a great Square in the Field; from the Sides of which the Off-sets are taken.

I have drawn this following Figure so, that at once you may see all the variety of this way of Working. The best way indeed is to contrive your Square so, that if possible, you may from the Sides thereof go upon a perpendicular-line to any of the Angles. But if that cannot be, then Perpendicular-lines to the Sides may do as well, as you see here, 1, 5, 7, 6,

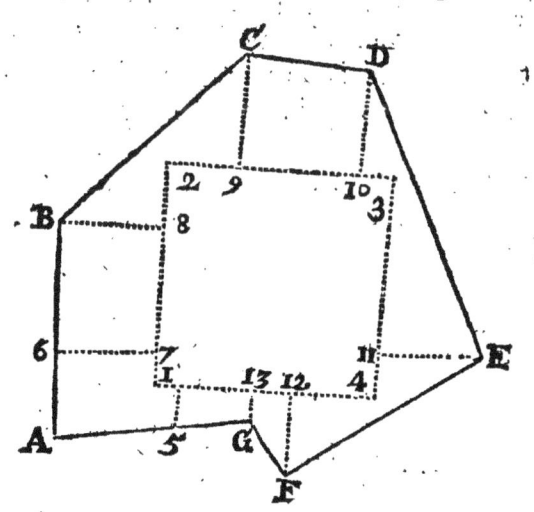

Hedge A C what number of Chains you please, no matter whether they be equal to the former or not; as A· 3 two Chains; next measure the distance 2, 3, *viz.* 1 Chain 68 Links; and then have you done in the Field. To plot which, draw the Line A B at adventure, and set off 2 Chains from A to 2; then take with your Compasses the distance A 3, 2 Chains, and setting one Foot in A, describe the Arch 2, 3, take also with your Compasses the distance 2, 3, *viz.* 1 Chain, 68 Links, and setting one Foot in 2. with the other cross the former Arch; through which cross draw the Line A C; which with A B will make an Angle equal to the Angle in the Field.

But the more easie and speedy way is to take but one Chain only along the Hedges; as in the foregoing Figure, I set a strong Stick in the very Angle A, and putting the Ring at one end of the Chain over it, I take the other end in my Hand, and stretch out the Chain along the first Hedge A B, and where it ends, as at 5, I stick down a stick, then I stretch the Chain also along the Hedge A C, and at the end thereof set another Stick as at 4, then loosing my Chain from A, I measure the distance 4, 5, which is 74 Links, which is all I need note down in my Field-Book for that Angle; and now coming to plot that Angle, I take first from my Scale the distance of one Chain, and placing one Foot of the Compasses in any part of the Paper, as at A, I describe the Arch 4, 5,; then I take from the same Scale 74 Links, and set it off upon that Arch, making marks where the ends of the Compasses fall, as at 4 5. Lastly, from A. through these Marks I draw the Line A B, and A C, which constitute the former Angle: Remember to
plot

Divers ways to take the Plots of Fields.

plot your Angles with a very large Scale; and you may set off your Lines with a smaller.

I will give you two *Examples* of this way of measuring, and then leave you to your own practice. First,

How by the Chain only to Survey a Field by going round the same.

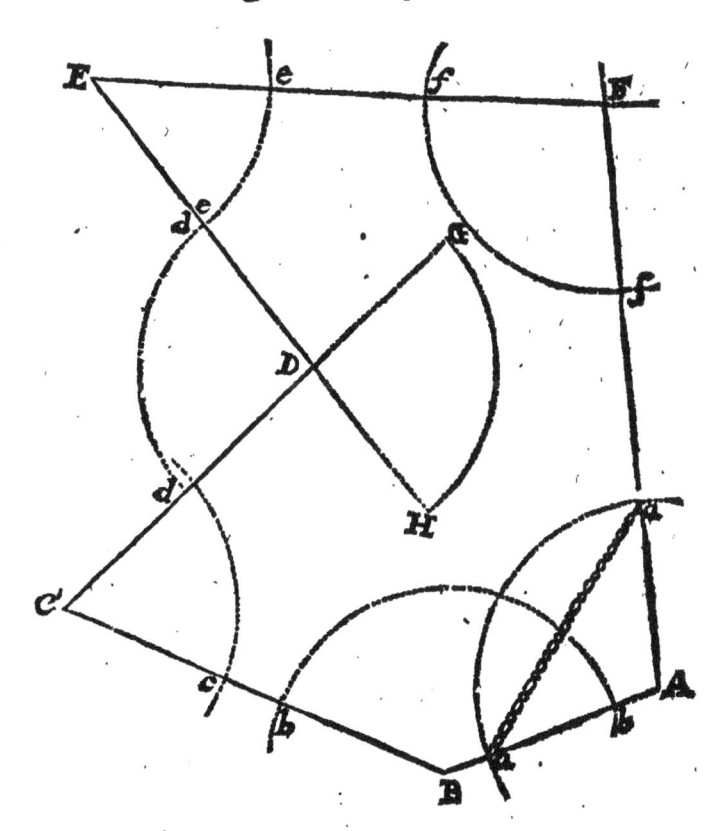

Let ABCDEF be the Field; and beginning at A in the very Angle, stick down a Staff through the

great

great Ring at one of the Ends of your Chain, and taking the other End in your Hand, stretch out the Chain in length, and see in what part of the Hedge A F the other End falls: as suppose at *a*, there let up a Stick; and do the like by the Hedge A B, and say there the Chain ends at (*a*) also; measure the nearest distance between *a* and *a*, which let be 1 Chain 60 Links, this note-down in your Field-Book; measure next the length of the Hedge A B, which is 12 Chains 50 Links; note this down also in your Field-Book. Nextly, coming to B, take that Angle in like manner as you did the Angle A, and measure the distance B C: after this manner you must take all the Angles, and measure all the Sides round the Field. But lest you be at a Nonplus at D, because this is an outward Angle, thus you must do; stick a staff down with the ring of the Chain round it in the very Angle D, then taking the other end of the Chain in your Hand, and stretching it at length, move your self to and fro, till you perceive your self in a direct line with the Hedge D C, which will be at G, where stick down an Arrow, or one of your Surveying-Sticks; then move round till you find your self in a direct line with the Hedge D E, and there the Chain stretched out at length, plant another Stick as, at H, then measure the nearest Distance, H G, which let be 1 Chain 43 Links; which note down in your Field-Book, and proceed on to measure the line D E; but in your Field-Book make some mark against D, to signifie it is an outward Angle, as >, or the like: And when you come to plot this, you must plot the same Angle outward that you took inward; for the Angle G D H, is the same as the Angle *d* D *d*. I made this outward

Angle

Divers ways to take the Plots of Fields. 115

Angle here on purpose to shew you how you must Survey a Wood, by going round it on the Outside, where you must take most of the Angles, as here you do D.

Having thus taken all the Angles, and measured all the Sides; the next thing to be done, is to lay down upon Paper, according to your Field Book: Which you will find to stand thus.

Angles	Cross Lines or Chords } Links Chains	Lines of the Field } Chains Links
A . 1 . 60	AB . 12 . 50
B . 1 . 84	BC . 23 . 37
C . 1 . 06	CD . 19 . 30
D . 1 . 43	DE . 20 . 00
E . 0 . 80	EF . 29 . 00
F . 1 . 52	FA . 31 . 50

Forasmuch now as it is convenient that the Angles be made by a greater Scale than the lines are laid down with: I have therefore in this Figure made the Angles by a Scale of one Chain in an Inch, and laid down the lines by a Scale of ten Chains in one Inch. But to begin to plot, take from your Scale one Chain, and with that distance, in any convenient place of your Paper, as at A, sweep the Arch *a, a*; then from the same large Scale, take off 1 Chain 60 Links, and set it upon that Arch, as from *a* to *a*; and from A draw the lines through *a* and *a*, as the lines AB, AF.

I 2 Then

116　*Divers ways to take the Plots of Fields.*

Then repairing to your shorter Scale, take from thence the first Distance, *viz.* 12 Chains 50 Links, and set it from A to B, drawing the Line A B.

Secondly, Repairing to B, take from your large Scale 1 Chain and setting one Foot of the Compasses in B, with the other make the Arch *b b*; also from the same Scale take your Chord-line, *viz.* 1 Chain 84 Links, and set it upon the Arch *b b*, one Foot of the Compasses standing where the Arch intersects A B, the other will fall at *b*; then through *b* draw the line B C; and from your smaller Scale set off the Distance B C 23 Chains, 37 Links, which will fall at C, where the next Angle must be made. After this manner proceed on according to your Field-Book, till you have done.

And here mark, that you need neither in the Field, nor upon the Paper, take notice of the Angle F, nor yet measure the lines E F and A F, for if you draw those two Lines through, they will intersect each other at the true Angle F: However, for the Proof of your Work, it is good to measure them, and also to take the Angle in the Field.

I must not let slip in this Place the usual way taught by Surveyors, for the measuring a Field by the Chain only, as true indeed as the former, but more tedious, which take as followeth.

The common way taught by Surveyors for taking the Plot of the foregoing Field.

Because I will not confound your Understanding with many lines in one Figure, I have here again placed the same. First, measure round the Field and
note

Divers ways to take the Plots of Fields. 117

note down in your Field-Book every Line thereof, as in this Field has been before done.

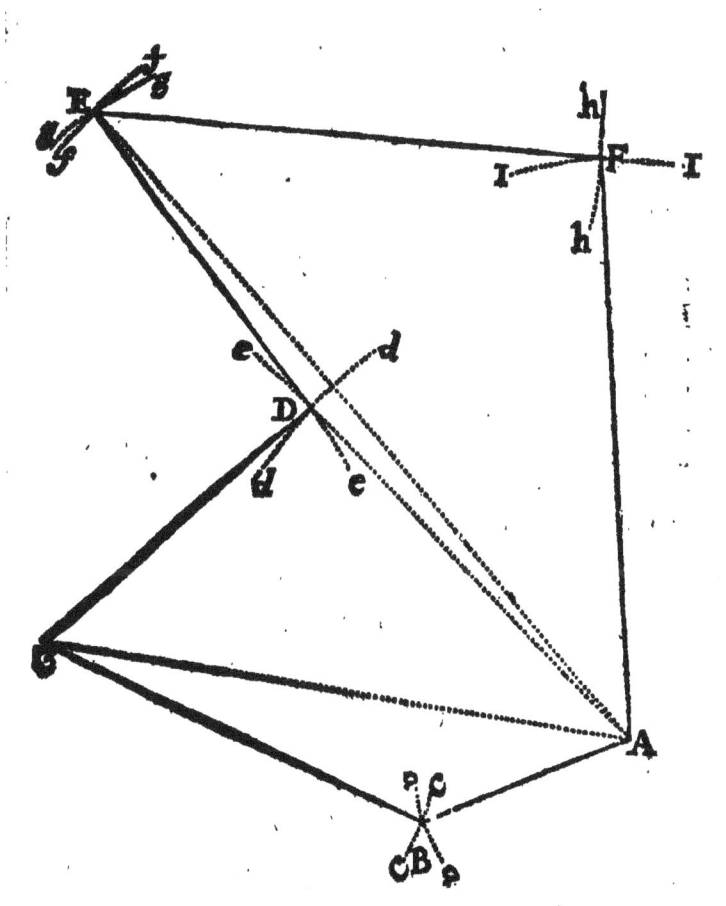

Secondly, turn all the Field into Triangles, as beginning at A, to measure the Diagonal A C, A D, A E, and note them down; then is your Field turned into four Triangles, and the Diagonals are

AC

118 *Divers ways to take the Plots of Fields.*

	Chains	Links
AC	33	70
AD	25	70
AE	45	40

To plots which, first draw a Line at adventure, as the line A C, and set off thereon 33 Chains 70 Links, according to your Field-Book for the Diagonals; then taken with your Compasses the length of the line A B *viz.* 12 Chains 50 Links, set one Foot in A, and with the other describe the Arch *a a*; also take the Line B C, *viz.* 23 Chains 37 Links, and setting one foot in C, with the other describe the Arch *c c*, cutting the Arch *a a* in the Point B, then draw the lines A B, C B, which shall be the two Bound-lines of the Field.

Secondly, Take with your Compasses the length of the Diagonal A D, *viz.* 25 Chains 70 Links, and setting one foot of the Compasses in A, with the other describe the Arch, as *d d*, also taking the line C D, *viz.* 19 Chains 30 Links, set one Foot in C, and with the other describe the Arch, *e e*, cutting the Arch *d d* in the point D, to which Intersection draw the line C D.

Thirdly, Take with your Compasses the length of the Diagonal A E, *viz* 45 Chains 40 Links, and setting one foot in A, with the other describe an Arch, as *f f* also take the line D E 20 Chains, and therewith cross the former Arch in the Point E, to which draw the line D E.

Lastly

Divers ways to take the Plots of Fields. 119

Laſtly, take with your Compaſſes the length of the line A F, *viz.* 31 Chains 50 Links; and ſetting one foot in A, deſcrbe an Arch, as I I. Alſo take the length of the line E F, *viz.* 29 Chains 00 Links, and therewith deſcribe the Arch *h h*, which cuts the Arch I I, in the Point F, to which Point draw the lines A F, and E F, and ſo will you have a true Figure of the Field.

I have ſhewed you both ways, that you may take your Choice. And now I proceed to my Second Example promiſed.

How to take the Plot of a Field at one Station, near the Middle therof, by the Chain only.

Let A B C D E be the Field, ☉ the appointed place, from whence by the Chain to take the Plot thereof. Stick a Stake up at ☉ through one ring of the Chain, and make your Aſſiſtant take the other end, and ſtretch it out. Then cauſe him to move up and down, till you eſpy him exactly in a line between the Stick and the Angle A; there let him ſet down a Stick, as at *a*, and beſure that the ſtick *a* be in a direct line between ☉ and A; which you may eaſily perceive by ſtanding at ☉, and looking to A. This done, cauſe him to move round towards B; and at the Chain's end, let him there ſtick down another ſtick exactly in the line between ☉ and B, as at *b*. Afterwards let him do the ſame at *c*, at *d*, and at *e*; and if there were more Angles, let him plant a ſtick at the end of the Chain in a right Line

I 4 between

between ⊙ and every Angle. In the next place measure the nigheſt diſtance between ſtick and ſtick, as *a b*, 1 Chain, 26 Links, *b c* 1 Chain 06 Links,

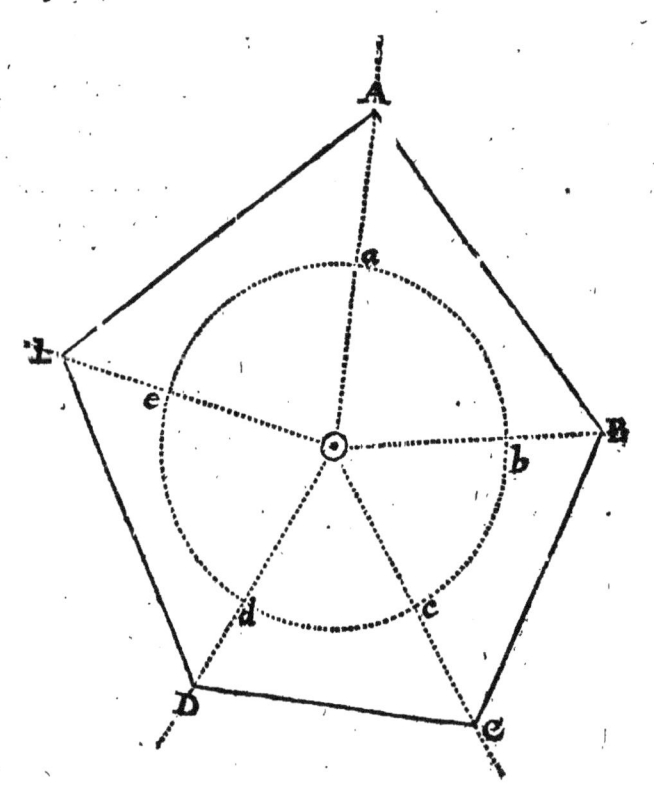

c d 1 Chain 00 Links, *d e*, 1 Chain 20 Links, and put them down in your Field-Book accordingly. Meaſure alſo the diſtances between ⊙ and every Angle, as ⊙ A 18 Chains 10 Links, ⊙ B 15 Chains 00 Links, &c. all which put down, your Field-Book will appear thus.

Subten-

Divers ways to take the Plots of Feilds.

Subtendent or chord-Lines.
$\begin{cases} ab & 1.26 \\ bc & 1.06 \\ cd & 1.00 \\ de & 1.20 \end{cases}$ Chains . Links

Diagonal or Centre-Lines
$\begin{cases} \odot A, & 18.10 \\ \odot B, & 15.00 \\ \odot C, & 17.00 \\ \odot D, & 15.00 \\ \odot E, & 16.00 \end{cases}$ Chains . Links

How to Plot the former Observations.

Take from a large Scale 1 Chain, and setting one foot of the Compasses in any convenient place of the Paper, as at ⊙, make the Circle *a b c d e*. Then taking for your first Subtendent, or Chord-line, 1 Chain 26 Links, set it upon the Circle, as from *a* to *b*. From ⊙ through *a* and *b*, draw lines, as ⊙ A, ⊙ B, which be sure let be long enough. Then take your second Subtendent from the same large Scale, *viz.* 1 Chain 6 Links, and set it upon the Circle from *b* to *c*, and through *c* draw the line ⊙ C. When thus you have set off all your Subtendents, and drawn lines through their several Marks, repair to a smaller Scale; and upon the lines drawn, set off your Diagonal or Centre-lines, as you find them in the Field-Book: So upon the line ⊙ *a* you must set off 18 Chains, 10 Links, making a Mark where it falls, as at A: Upon the line ⊙ *b* 15 Chains, 00 Links, which falls at B, and so by all the rest. Lastly, draw the lines AB, BC, CD, &c. and the Work will be finished.

It would be but running things over again to shew you how after this manner, to Survey a Field at two or three Stations, or in any Angle thereof, &c. For if you well understand this, you cannot be ignorant of the rest.

How

CHAP. VII.

How to cast up the Contents of a Plot of Land.

Having by this time sufficiently shewed you how to Survey a Field, and lay down a true Figure thereof upon Paper; I come in the next place to teach you how to cast up the Contents thereof; that is to say, to find out how many Acres, Roods and Perches it containeth. And first

Of the Square, and Parallelogram.

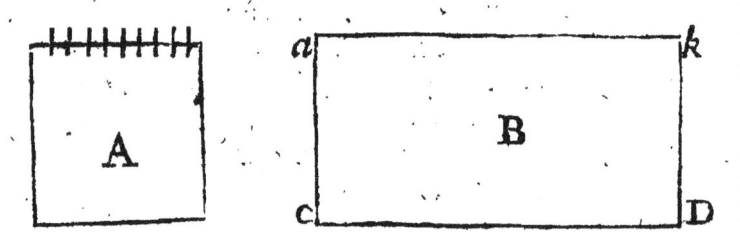

To cast up either of which, multiply one Side by the other, and the Product will be the Content.

EXAMPLE,

Let A be a true Square, each side being 10 Chains; multiply 10 Chains 00 Links by 10 Chains 00 Links, *facit* 10|00000, from which I cut off the five last Figures and there remains just 10 Acres for the Square A.

Again,

How to cast up the Contents of a Plot of Land. 123

Again, in the Parallelogram B, let the side *A b* or *c* D be 20 Chains, 50 Links; and the side *a c* or *b* D 10 Chains, 00 Links: Multiply *a b* 20 Chains, 50 Links, by *a c* 10 Chains, 00 Links, *facit* 20,50000. from which cutting off the last Figures, remains 20 Acres. Then if you multiply the Figures cut off, *viz.* 50000 by 4, *facit* 200000; from which cutting off five Figures, remains 2 Roods; and if any thing but 000 had been left, you must have multiplied again by 40; and then cutting off again five Figures, you would have had the odd Perches: See it done hereunder.

I need not have multiplied 00 by 40; for I know 40 times Nothing is Nothing; but only to shew you in what order the Figures will stand when you have odd Perches, as presently we shall light on. So much is the Content of the long Square B, *viz.* 20 Acres, 2 Roods, 00 Perch.

	20 50
	10.00
Acres	20|50000
	4
Roods	2|00000
	40
Perches	0 000 00

Of Triangles.

The Content of all Triangles are found, by multiplying half the Base by the whole Perpendicular; or the whole Base by half the Perpendicular; or otherwise, by multiplying the whole Base and whole Perpendicular together, and taking half that Product for the Content. Either of these three ways will do, take which you please.

E X A M-

EXAMPLE.

In the Triangle A, the Base *a b* is 10 Chains 00 Links: the Perpendicular *c b* 13 Chains, 70 Links: the half of which is 6 Chains, 85 Links; which multiplied by 10 Ch. 00 Lin. *facit* 685000; from which cutting off five Figures, there is left 6 Acres. Then multiplying the Remainder by 4, *facit* 340000; from which taking five Figures, remains 3 Roods. Again the five Figures cut off multiplied by 40, makes 1600000; from which taking five Figures, leaves 16 Perches. See the Operation.

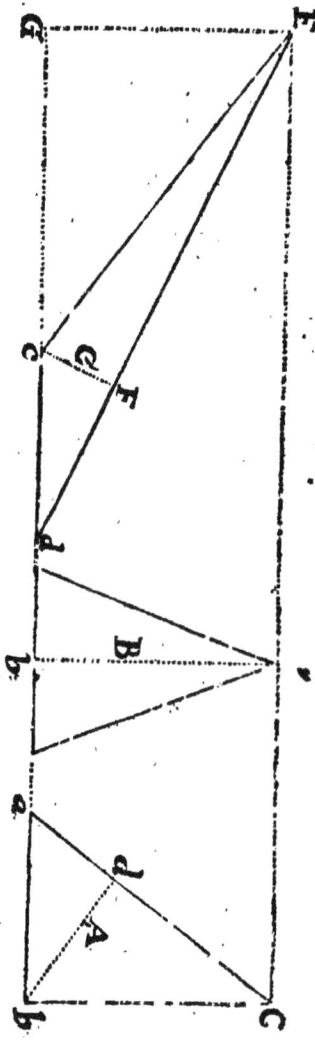

```
            6, 85
           10,00
          ────────
Acres    6|85000
               4
          ────────
Roods    3|40000
              40
          ────────
Perches 16|00000
          ────────
```

How to cast up the Contents of a Plot of Land. 125

So likewise in the Triangle B, the Perpendicular *a b* is 13 Chains, 70 Links, which multiplied by half the Base, will give the same Content.

Also in the Triangle C, if you multiply half the Base E *d*, by the Perpendicular *c* F, the Product will be the Content of the Triangle.

And here Note, that you are not confined to any Angle, but you may let fall your Prependicular from what Angle you please, taking the Line on which it falls for the Base. Thus in the Triangle A, if from *b* you let fall a Perpendicular, take *b*, *d*, and the half of *a c* for finding the Content. Also in the Triangle C, you may from E let fall your Perpendicular, although it falls without the Triangle; and the half of E G, and the whole of *c d*, shall be the true Content of the Triangle C; but then you must remember to extend the Base-line *c d*.

Remember this, all Triangles having equal Bases and lying between Parallel-lines, are of the same Content; so the Triangles A B C having equal Bases and lying between the lines E C and G *b*, are therefore of the same Content.

To find the Content of a Trapezia.

Draw between two opposite Angles a streight Line, as A B; then is the Trapezia reduced into two Triangles, *viz.* ABC and ABD, which you may measure as before taught, and adding their Products together, you will have the true content of the Trapezia. Or a little shorter, thus:

Take

126 How to cast up the Contents of a Plot of Land.

Take the length of the line A B, which let be 37 Chain 00 Links; take also the length of the Perpendicular D e, which let be 7 Chains 40 Links; also C d 4 Chains 80 Links; add the two Perpendiculars together, and they make 12 Chains 20 Links, which multiply by half the common Base A B 18 Chains 50 Links, and the Product is 22 Acres, 2 Rood, 11 Perch, as appears by the Operation hereunder.

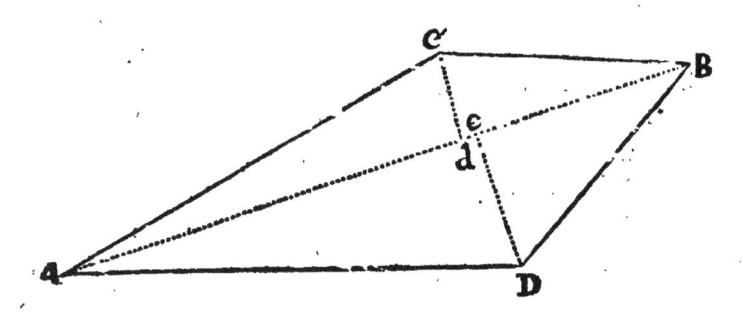

    Half the common Base A B   18,50
    The Sum of the two Perpendiculars   12,20

                                  37000
                                   3700
                                   1850

                 Acres   22|5 7000
                                         4

                 Roods   2|2 8000
                                      40

                 Perches   11|20000

How

How to cast up the Contents of a Plot of Land. 127

How to find the Content of an Irregular Plot, consisting of many Sides and Angles.

To do this, you must first by drawing Lines from Angle to Angle, reduce all the Plot into Trapezia's and Triangles, after which measure every Trapezia and Triangle severally, and adding their Contents all together, you will have the true Content of the whole Plot.

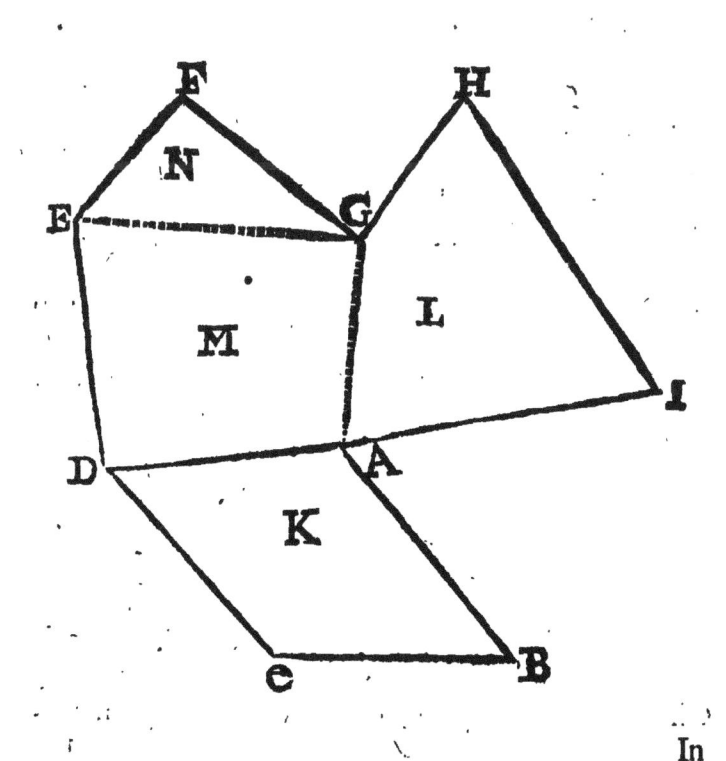

In

128 *How to cast up the Contents of a Plot of Land.*

In the annexed Figure A B C D E F G H I, draw the Line A D, which cuts off the Trapezia K; also the line A G, which cuts off the Trapezia L: And lastly the line G E, which makes the Trapezia M, and the Triangle N, so is the whole Plot reduced into the three Trapezia's K, L, M, and the Triangle N; all which I measure as before taught, and put them down as hereunder.

	Acres	Roods	Perches
The Trapezia K contains	21	2	12
The Trapezia L contains	26	3	18
The Trapezia M contains	30	2	16
The Triangle N contains	6	2	24
The Content of the Plot	85	2	30

By which you find the whole Plot to contain 85 Acres, 2 Rood, 30 Perches.

If the sides of the Plot had been given in Perches, Yards, Feet, or any other Measure, you must still cast up the Content after this manner, and then your Product will be Perches, Yards, &c. To turn which into Acres, Roods and Perches, I have largely treated of in the beginning of this Book.

How to find the Content of a Circle, or any Portion thereof.

To find the Content of the whole Circle, it is convenient, that first you know the Diameter and Circumference thereof; one of which being known, the

How to cast up the Contents of a Plot of Land. 129

the other is easily found; for as 7 is to 22, so is the Diameter to the Circumference. And as 22 is to 7 so is the Circumference to the Diameter.

In this annexed Figure, the Diameter A B is 2 Chains, or 200 Links, which multiplied by 22, and

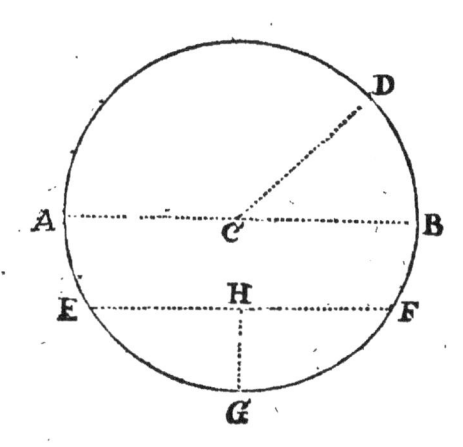

the Product divided by 7, gives 6 Chains 28 Links, and something more for the Circumference. Now, to know the Superficial Content multiply half the Circumference by half the Diameter, the Product will be the Content: Half the Circumference is 3 Chains 14 Links: half the Diameter 1 Chain 00 Links; which multiplied together, the Product is 3,1400 square Links, or 1 Rood 10 Perch, the Content of the Circle. Again,

By the Diameter only find the Content.

As 14 is to 11, so is the square of the Diameter to the Content. The square of the Diameter is 40000,

K which

which multiplied by 11, makes 440000, which divided by 14, gives 31428, or 1 Rood 14 Perch, and something more for the Content.

How to measure the Superficial Content of the Section of a Circle.

Multiply half the Compass thereof by the Semidiameter of the Circle, the Product will answer your desire.

In the foregoing Circle, I would know the Content of that little piece DCB; the Arch D B is 78 Links $\frac{1}{2}$; the half of it 39 $\frac{1}{4}$, which multiplied by 1 Chain, 00 Links, the Semi-diameter gives 3925 Square Links, or 6 $\frac{1}{4}$ perches.

How to find the Content of a Segment of a Circle without knowing the Diameter.

Let EFG be the Segment, the Chord E F is 1 Chain 70 Links, or 170 Links, the Perpendicular G H 50 Links; now multiply $\frac{2}{3}$ of the one by the whole of the other, the Product will be the Content, the two Thirds of 170 is nearest 113, which multiplied by 50, produces 5650 square Links, or 9 Perches.

How to find the Superficial Content of an Oval.

The common way is to multiply the long Diameter by the shorter, and observe the Product and then

How to cast up the Contents of a Plot of Land. 131

then as if you where measuring a Circle, say,

As 14 to 11, so the said Product to the Content of the Oval; but this is not exact: A better way is;

As 1, 11/100 is to the length of the Oval; so is the breadth to the Content, or nearer, as 1,27324 to the length; so the breadth to the Content.

How to find the Superficial Content of Regular Polygons; as Pentagons, Hexagons, Heptagons, &c.

Multiply half the summ of the Sides, by a Perpendicular, let fall from Centre upon one of the Sides, the Product will be the Area or Superficial Content of the Polygon. In the following Pentagon the side BC is 84 Links, the whole Summ of the five Sides, therefore must be 420, the half of which is 210, which multiplied by the Perpendicular AD 56 Links gives 11760 Square Links for the Content, or 18 Perches 6/10 of a Perch, almost 19 Perches

I have been shorter about these three last Figures than my usual Method because they very rarely fall into the Surveyors way to measure them in Land, though indeed in Broad Measure, Paving, &c. often.

K 2 CHAP

CHAP. VIII.

Of laying out New Lands, very useful for Surveyors, in His Majesty's Plantations in America.

A certain quantity of Acres being given, how to lay out the same in a Square Figure.

ANnex to the number of Acres given, 5 Cyphers, which will turn the Acres into Links; then from the Number thus increased, extract the Root, which shall be the Side of the proposed Square.

EXAMPLE.

Suppose the Number given be 100 Acres which I am to lay out in a square Figure; I join to the 100 5 Cyphers, and then it is 100,00000 square Links, the Root of which is 3162 nearest, or 31 Chains 62 Links, the length of one side of the Square.

Again,

If I were to cut out of a Corn-Field one square Acre; I add to one five Cyphers, and then is it 100000, the Root of which is 3 Chains 16 Links, and something more for the side of that Acre.

How

How to lay out any given quantity of Acres in a Parallelogram; whereof one Side is given.

Turn first the Acres into Links, by adding as before, 5 Cyphers, that number thus increased, divide by the given Side, the Quotient will be the other side.

EXAMPLE.

It is required to lay out 100 Acres in a Parallelogram, one Side of which shall be 20 Chains, 00 Links; first to the 100 Acres I add 5 Cyphers, and it is 100,00000; which I divide by 20 Chains 00 Links, the Quotent is 50 Chains 00 Links, for the other side of the Parallelogram.

How to lay out a Parallelogram that shall be 4, 5 6, or 7, &c. times longer then it is broad.

In *Carolina*, all Lands lying by the sides of Rivers, except Seigniories or Baronies, are (or ought, by Order of the Lord's Proprietors to be) thus laid out. To do which, first as above taught, turn the given quantity of Acres into Links, by annexing 5 Cyphers; which summ divide by the number given for the Proportion between the length and breadth, as 4, 5, 6, 7, &c. the Root of the Quotient will shew the shortest Side of such a Paralle logram.

EXAMPLE

EXAMPLE.

Admit it were required of me to lay out 100 Acres in a Parallelogram, that should be five times as long as broad: First to the 100 Acres I add 5 Cyphers, and it makes, 100,00000, which summ I divide by 5, the Quotient is 2000000, the Root of which is nearest 14 Chains 14 Links, and that I say shall be the short Side of such a Parallelogram, and by multiplying that 1414 by 5, shews me the longest Side thereof to be 70 Chains 70 Links.

How to make a Triangle that shall contain any number of Acres, being confined to a certain Base

Double the given number of Acres, (to which annexing first five Cyhers) divide by the Base; the Quotient will be the length of the Perpendicular.

EXAMPLE

Upon a Base given that is in length 40 Chains, 00 Links; I am to make a Triangle that shall contain 100 Acres. First I double the 100 Acres, and annexing five Cyphers thereto, it makes 200,00000. which I divide by 40 Chains, 00 Links the limited Base; the Quotient is 50 Chains 00 Links for the height of the Perpendicular. As in this Figure, A B is the given Base 40; upon any part of which Base, I set the Perpendicular 50, as at C; then the Perpendicular is C D. Therefore I draw the Lines DA, DB, which

Divers ways to take the Plots of Fields. 135

which makes the Triangle D A B to contain juſt 100 Acres, as required. Or if I had ſet the Perpendicular at E, then would EF have been the Perpendicular 50, and by drawing the Lines F A, F B; I ſhould have made the Triangle F A B, containing 100 Acres, the ſame as D A B.

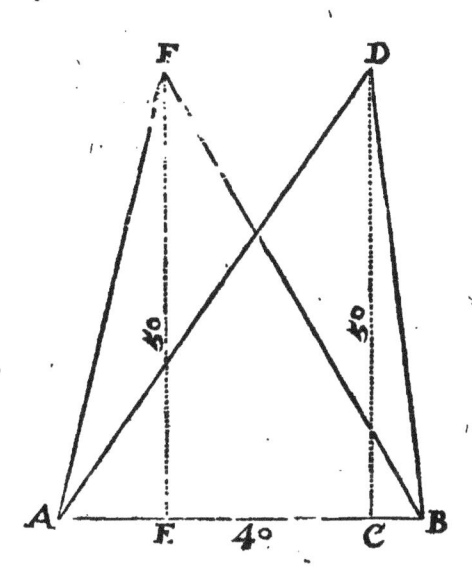

If you conſider this well when you are laying out a new peice of Land, of any given content, in *America*, or elſewhere, although you meet in your way with 100 Lines and Angles; yet you may, by making a Triangle to the firſt Station you began at, cut off any quantity required.

How to find the Length of the Diameter of a Circle which shall contain any number of Acres required.

Say as 11 is to 14, so will the number of Acres given be to the Square of the Diameter of the Circle required,

EXAMPLE.

What is the length of the Diameter of a Circle, whose Superficial Content shall be 100 Acres? Add five Cyphers to the 100, and it makes 100,00000 Links, which multiply by 14, *facit* 140000000 ; which divided by 11, gives for Quotient 12727272, the Root of which is 35 Chains, 67 Links and better, almost 68 Links. And so much shall be the Diameter of the required Circle.

I might add many more Examples of this Nature, as how to make Ovals, Regular Polygons, and the like, that should contain any assigned quantity of Land. But because such things are meerly for Speculation, and seldom or never come in Practice, I at present omit them,

CHAP.

CHAP. IX.

Of Reduction.

How to Reduce a large Plot of Land or Map into a lesser compass, according to any given Proportion; or e contra how to Enlarge one.

THE best way to do this, is, if your Plot be not over-large, to plat it over again by a smaller Scale: But if it be large, as the Map of a County, or the like, the only way is to compass in the Plot first with one great Square; and afterwards to divide that into as many little Squares as you shall see convenient. Also make the same Number of little Squares upon a fair Peice of Paper, by a lesser Scale, according to the Proportion given. This done, see in what Square, and part of the same Square, any remarkable Accident falls, and accordingly put it down in your lesser Squares; and that you may not mistake, it is a good way to number your Squares. I cannot make it plainer, than by giving you the following Example, where the Plot ABCD, made by a Scale of 10 Chains in an Inch, is reduced into the Plot EFGH, of 30 Chains in an Inch.

There

138 Reduction of Land.

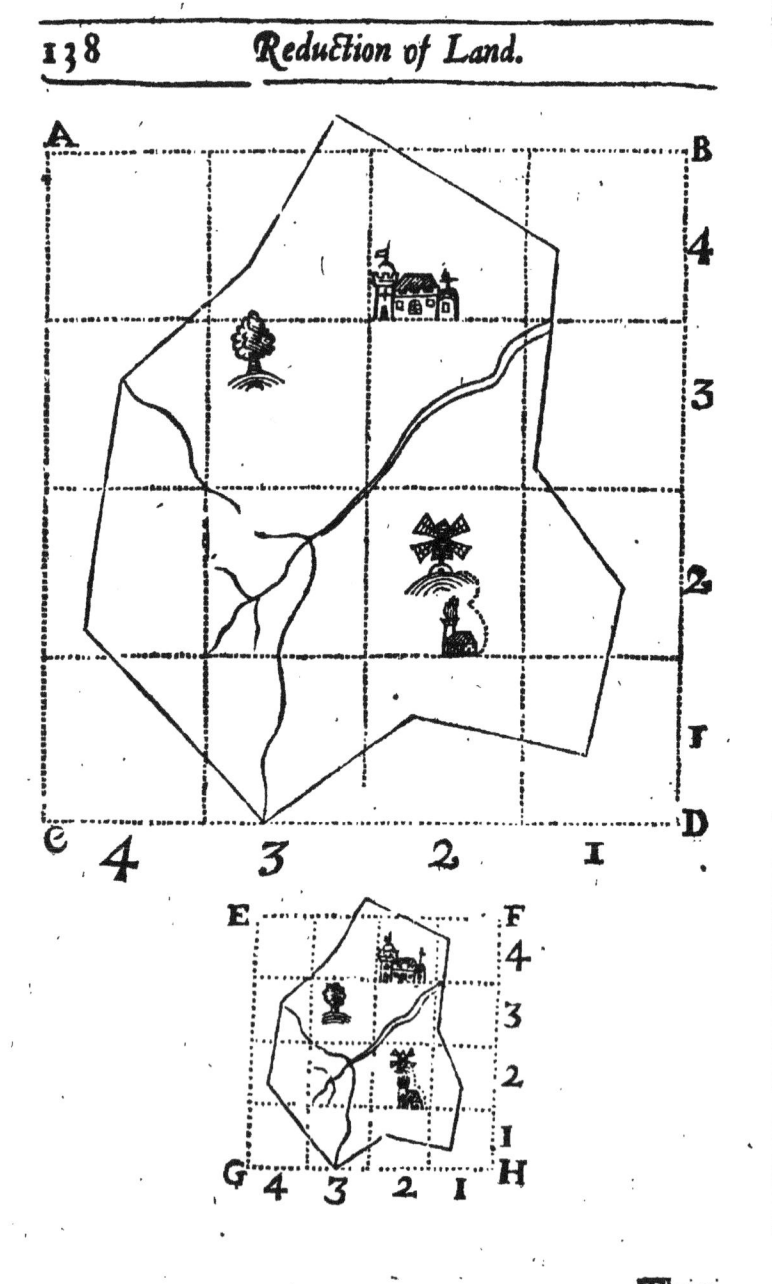

There

Reduction of Land   139

   There are several other ways taught by Surveyors for reducing Plots or Maps, as Mr. *Rathborn*, and after him, Mr. *Holwell* adviseth to make use of a Scale or Ruler; having a Centre hole at one end, thro' which to fasten it down on a Table, so that it may play freely round, and numbred from the Centre-end to the other, with lines of Equal Parts: The Use of which is thus. Lay down upon a smooth Table, the Map or Plot that you would reduce and glew it with Mouth-glew fast to the Table at the four corners thereof. Then taking a fair piece of Paper about the Bigness that you would have your reduced Plot to be of, and lay that down upon the other; the middle of the last about the middle of the first. This done, lay the Centre of your Reducing-Scale near the Centre of the white Paper, and there with a Needle through the Centre make it fast; yet so, that it may play easily round the Needle. Then moving your Scale to any remarkable thing of the first Plot, as an Angle, a House, the bent of a River, or the like: See against how many Equal Parts of the Scale it stands, as suppose 100; then taking the $\frac{1}{2}$, the $\frac{1}{4}$, the $\frac{1}{3}$, or any other number thereof according to the proportion you would have the reduced Plot to bear, and make a mark upon the white Paper against 50, 25, 33, &c. of the same Scale: And thus turning the Scale about, you may first reduce all the outermost parts of the Plot. Which done you must double the lesser Plot, first $\frac{1}{2}$ thereof, and then the other; by which you may see to reduce the innermost part near the Centre.

   But I advise rather to have a long Scale, made with the Centre-hole, for fixing it to the Table in about one third part of the Scale, so that $\frac{2}{3}$ of the

Scale

Scale may be one way numbred with Equal Parts from the Centre-hole to the end; and ½ part thereof numbred the other way to the end with the same number of Equal Parts, tho lesser. Upon this Scale may be several Lines of Equal Parts, the lesser to the greater, according to several Proportions. Being thus provided with a Scale, glew down upon a smooth Table your greater Plot to be reduced; and close to it upon the same Table, a Paper, about the bigness whereof you would have your smaller Plot. Fix with a strong Needle the Centre of your Scale between both; then turning the longer end of your Scale to any remarkable thing of your Plot, to be reduced see what number of Equal Parts it cuts, as suppose 100; there holding fast the Scale, against 100 upon the smaller end of your Scale, make a mark upon the white Paper; so do round all the Plot, drawing lines, and putting down all other accidents as you proceed, for fear of Confusion through many Mark in the end; and when you have done although at first the reduced Plot will seem to be quite contrary to the other; yet when you have unglewed it from the Table and turned it about, you will find it to be an exact Epitome of the first. You may have for this Work divers Centres made in one Scale, with equal Parts proceeding from them accordingly; or you may have divers Scales, according to several Proportions, which is better.

What has been hitherto said concerning the Reducing of a Plot from a greater Volume to a lesser, the same is to be understood *vice versa*, of Enlarging Plot, from a lesser to a greater. But this last seldom comes in practise.

Reduction of Land. 141

How to change Customary-Measure into Statute, and the contrary.

In some parts of *England*, for Wood-Lands; and in most parts of *Ireland*, for all sorts of Lands; they account 18 Foot to a Perch, and 160 such Perches to make an Acre, which is called *Customary-Measure* Whereas, our true Measure for Land, by *Act of Parliament*, is but 160 Perches for one Acre, at 16 Foot ½ to the Perch. Therefore to reduce the one into the other, the Rule is,

    As the Square of one sort of Measure,
    is to the Square of the other:
    So is the content of the one,
    to the Content of the other.

Thus, if a Field measured by a Perch of 18 Feet, accounting 160 Perches to the Acre, contain 100 Acres; How many Acres shall the same Field contain by a Perch of 16 Feet ½ ?

Say, if the Square of 16 Feet ½ *viz.* 272.25. give the Square of 18, Feet *viz.* 324. what shall 100 Acres *Customary* give? Answer 119 ⅘ of an Acre Statute.

Knowing the Content of a Piece of Land, to find out what Scale it was Plotted by.

First, By any Scale measure the Content of the Plt which done, argue thus:

    As the Content found, is to the Square of the
      Scale I tryed by;
    So is the true Content, to the Square of the true
      Scale it was plotted by.

                              Admit

Admit there is a Plot of a piece of Land containing 10 Acres, and I meafuring it by the Scale 11 in an Inch, find it to contain 12 Acres $\frac{1}{10}$ of an Acre Then I fay, If 12 $\frac{1}{10}$ give for its Scale 11 : What fhall 100 give? Anfwer 10. Therefore I conclude that Plot to be made by a Scale of 10 in the Inch. And fo much concerning Reducing Lands.

CHAP. X.

Inftructions for Surveying a Mannor, County or whole Country.

To Survey a Mannor, obferve thefe following RULES.

1. Walk or ride over the Mannor once or twice that you may have as it where a Map of it in your Head, by which means, you may the better know where to begin, and proceed on with your Work.

2. If you can conveniently run round the whole Mannor with your Chain and Inftrument, taking all the Angles, and meafuring all the Lines thereof; taking notice of Roads, Lanes or Commons, as you

crofs

cross them: Also minding well the Ends of all dividing Hedges, where they butt upon your Bound-Hedges in this manner.

3. Take a true Draught of all the Roads and By-Lanes in the Mannor, putting down also the true buttings of all the Field-Fences to the Road. If the Road be broad, or goes through some Common or Waste Ground, the best way is to measure, and take the Angles on both sides thereof; but if it be a narrow Lane, you may only measure along the midst thereof, taking the Angles and Off-sets to the Hedges, and measure your Distances truly: Also if there be any considerable River either bounds or runs thro' the Mannor, survey that also truly, as is hereafter taught.

4. Make a true Plot upon Paper of all the foregoing Work; and then will you have a Resemblance of the Mannor, though not compleat, which to make so, go to all the Buttings of the Hedges, and there, survey every Field distinctly, plotting it accordingly every Night, or rather twice a Day, till you have perfected the whole Mannor.

5. When thus you have plotted all the Fields, according to the Buttings of the Hedges found in your first Surveys, you will find that you have very nigh, if not quite done the whole Work: But if there be any Fields lie so within others, that they are not bounded on either side by a Road, Lane, nor River, then you must also survey them, and place them in your Plot, accordingly as they are bounded by other Fields.

6 Draw

6. Draw a fair Draught of the whole, putting down therein the Mannor-Houſe, and every other conſiderable Houſe, Wind-mill, Water-mill, Bridge, Wood, Coppice, Croſs-paths, Rills, Runs of Water, Ponds, and any other Matters Notable therein. Alſo in the fair Draught, let the Arms of the Lord of the Mannor be fairly drawn, and a Compaſs in ſome waſte part of the Paper; alſo a Scale, the ſame by which it was plotted: You muſt alſo beautify ſuch a Draught with Colours and Cuts according as you ſhall ſee convenient.

Write down alſo in every Field the true Content thereof; and if it be required, the Names of the preſent Poſſeſſors, and their Tenures; by which they hold it of the Lord of the Mannor.

The Quality alſo of the Land, you may take notice of as you paſs over it, if you have Judgment therein, and it be required of you.

How to take the Draught of a County or COUNTRY.

1. If the County or Country is in any place thereof bounded with the Sea, ſurvey firſt the Sea-coaſt thereof, meaſuring it all along with the Chain, and taking all the Angles thereof truly.

2. Which done, and plotted by a large Scale, ſurvey next all Rocks, Sands, or other Obſtacles that lie at the entrance of every River, Harbour, Bay or Road upon the Coaſt of that County, or Country; which plot down accordingly, as I ſhall teach you in this Book by and by.

3. Sur-

Instructions for Surveying a Mannor. 145

3. Survey all the Roads, taking notice as you go along of all Towns, Villages, great Houses, Rivers, Bridges, Mills, Cross-Ways, &c. Also take the bearing at two Stations of such Remarks, as you see out of the Road, or by the Side thereof.

4. Also Survey all the Rivers, taking notice how far they are Navigable, what (and where the) Branches runs into them, what Fords they have, Bridges, &c.

5. All this being exactly plotted, will give you a truer Map of the Country then any that I know of hath been yet made in *England*. However you may look upon old Maps, and if you find therein any thing worth the Notice that you have not yet put down, you may go and Survey it; and thus by degrees you may so finish a Country, that you need not so much as leave out one Gentleman's House; for hardly will it scape, but some very remarkable thing will come into your View, either from the Roads, the Rivers, or Sea-coast.

6. Lastly, with a large Quadrant take the true Latitude of the Place, in three or four Places of the County, which put down upon the Edge of your Map accordingly.

L   CHAP.

CHAP. XI.
Of dividing LANDS.

How to divide a Triangle several ways.

SUppose A B C to be a Triangular Piece of Land containing 60 Acres, to be divided between two Men, the one to have 40 Acres cut off towards A, and the other 20 Acres towards C; and the line of Division to proceed from the Angle B. First measure the Base A C, viz. 50 Chains, 00 Links; then say by the Rule of Three, If the whole Content 60 Acres give 50 Chain for its Base, what shall 40 Acres give? Multiply and Divide, the Quotient shall be 33 Chains 33 Links; which set off upon the Base from A to D, and draw the Line B D, which shall divide the Triangle as was required. If it had been required to have divided the same into 3, 4, 5, or more unequal Parts, you must in the like manner, by the Rule of Three have found the length of each several Base; much after the same manner as Merchants put their Gains By the Rule of Fellowship.

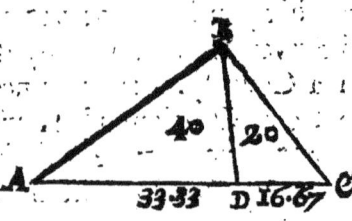

There are several ways of doing this by Geometry without the help of Arithmetick, but my Business is

not

Of Dividing Lands, 147

not to shew you what may be done, but to shew you how to do it, the most easie and practicable way.

How to divide a Triangular Piece of Land into any Number of Equal or Unequal Parts, by Lines proceding from any Point assigned, in any Side thereof.

Let A B C the Triangular Piece of Land, containing 60 Acres to be divided between three Men, the first to have 15 Acres, the second 20, and the third 25 Acres, and the Lines of Division to proceed from D: First measure the Base, which is 50 Chains; then divide the Base into three Parts, as you have been before taught, by saying, If 60 give 50, what shall 15 give? Answer, 12 Chains 50 Links for the first Man's Base; which set off from A to E. Again, say, If 60 give 50, what shall 20 give? Answer, 16 Chains 66 Links for the second Man's Base; which set off from E to F, then consequently the third Man's Base, viz. from F to C must be 20 Chains 84 Links: This done, draw an obscure Line from the Point assigned D, to the opposite Angle B, and from E and F draw the Lines EA and FG, parallel to BD. Lastly, from D, draw the Lines DH, DG, which shall divide the Triangle into three such Parts as was required.

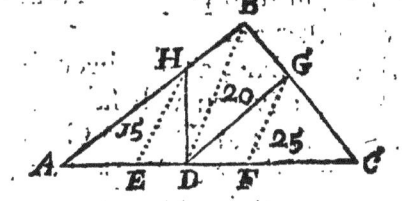

L 2 *How*

Of Dividing of Lands.

How to divide a Triangular Piece of Land, according to any Proportion given by a Line Parallel to one of the Sides.

A B C is the Triangular Piece of Land, containing 60 Acres, the Base A C is 50 Chains; this Piece of Land is to be divided between two Men, by a Line Parallel to B C, in such proportion that the one have 40 Acres, the other 20.

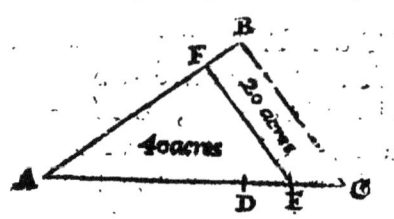

First Divide the Base, as has been before taught, and the point of Division will fall in D, A D being 33 Chains, 33 Links, and D C 16 Chains 67 Links.

Secondly, Find a mean Proportion between AD and A C; by multiplying the whole Base 50 by A D 33, 33, the Product is 16665000, of which summ extract the Root, which is 40 Chains 82 Links, which set off from A to E. Lastly, from E draw a Line parallel to B C, as is the Line E F; which divides the Triangle, as demanded.

Of Dividing Four-Sided Figures or Trapezia's.

Before I begin to teach you how to divide Pieces of Land of four Sides, it is convenient first to shew you how to change any Four-sided Figure into a Triangle; which

Of Dividing of Lands. 149

which done, the Work will be the same as in dividing Triangles.

How to reduce a Trapezia into a Triangle, by Lines drawn from any Angle thereof.

Let ABCD be the Trapezia to be reduced into a Triangle, and B the Angle assigned: Draw the

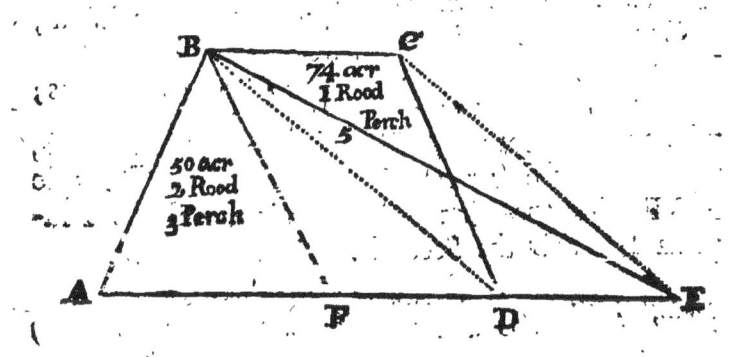

Dark Line BD, and from C make a Line Parallel thereto, as CE; extend also the Base AD, till it meet CE in E, then draw the Line BE, which shall make the Triangle BAE equal to the Trapezia ABCD.

Now to divide this Trapezia according to any assigned Proportion is no more but to divide the Triangle ABE, as before taught, which will also divide the Trapezia.

EXAMPLE.

Suppose the Trapezia ABCD containing 124 Acres 3 Roods and 8 Perches, is to be divided between two Men, the first to have 50 Acres, 2 Rood.

and

150 Of Dividing Lands.

and 3 Perches; the other 74 Acres, 1 Rood and 5 Perches, and the line of Division to proceed from B.

First, Reduce all Acres and Roods into Perches, then will the Content of the Trapezia be 19968 Perches; the first Man's Share 8083 Perches: the second 11885.

Secondly, Measure the Base of the Triangle, viz A E 78 Chains 00 Links;

Then say, If 19968 the whole Content give for its Base } 78 Chains 00 Links, What shall 8083, the first Man's part give? Answer } 31 Chains 52 Links; which set off from A to F, and drawing the Line FB, you divide the Trapezia as desired; the Triangle A B F being the First Man's Portion, and the Trapezia B C F D the Seconds.

How to reduce a Trapezia into a Triangle, by Lines drawn from a Point assigned in any Side therof.

A B C D the Trapezia, E the Point assigned, from whence to reduce it into a Triangle, and run the division Line; the Trapezia is of the same Con-

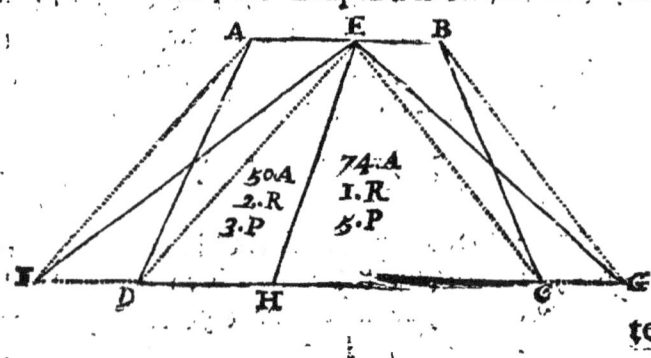

Of Dividing Lands. 151

tent as the former, *viz.* 19968 Perches, and it is to be divided as before, *viz.* one Man to have 8083 Perches, and the other 11885. First, for to reduce it into a Triangle, draw the Lines ED, EC, and from A and B makes Lines parallel to them, as A F, B G; then draw the Lines E G, E F, and the Triangle E F G will be equal to the Tarpezia ABCD; which is divided as before; for when you have found by the Rule of Proportion, What the first Man's Base must be *viz.* 31 Chains 52 Links, set it from F to H, and draw the Line H E, which shall divide the Trapezia according to the former Proportion.

How to reduce an Irregular Five-Sided Figure into a Triangle, and to divide the same.

Let A B C D E be the Five-sided Figure; to reduce which into a Triangle, draw the Lines A C,

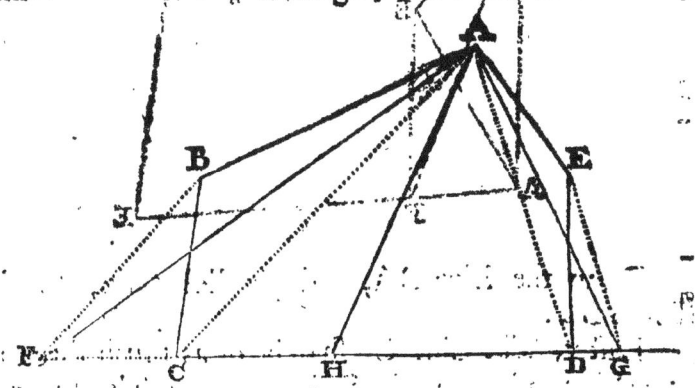

A D; and parallel there to B F, E G, extending the Base from C to F, and from D to G; then draw the Lines A F, A G, which will make the Triangle A F G equal to the Five-Sided Figure. If this was to

Of Dividing of Lands.

to be divided into two equal Parts, take the half of the Base of the Triangle, which is F H, and from H draw the Line H A; which divides the Figure A B C D E into two equal Parts. The like you may do for any other Proportion.

If in dividing the Plot of a Field there be Outward Angles, you may change them after the Following manner.

Suppose A B C D E be the plot of a Field; and B the outward Angle.

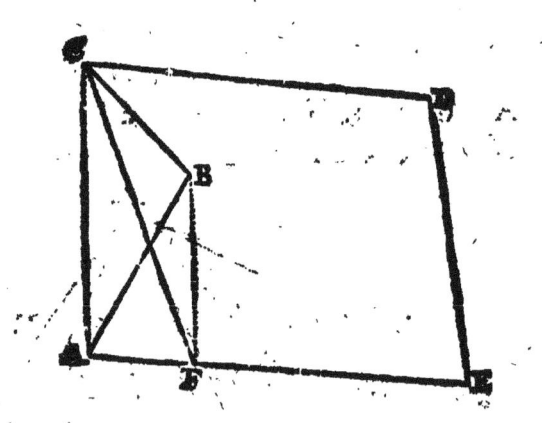

Draw the Line C A, and parallel thereunto the Line B F.

Lastly, The Line C F shall be of as much force as the Lines C B and B A. So is that Five-sided Figure, having one outward Angle reduced into a four-sided Figure, or *Trapezia*; which you may again reduce into a Triangle as has been before taught.

How

Of Dividing of Lands. 153

How to Divide an Irregular Plot of any number of Sides, according to any given Proportion, by a streight Line through it.

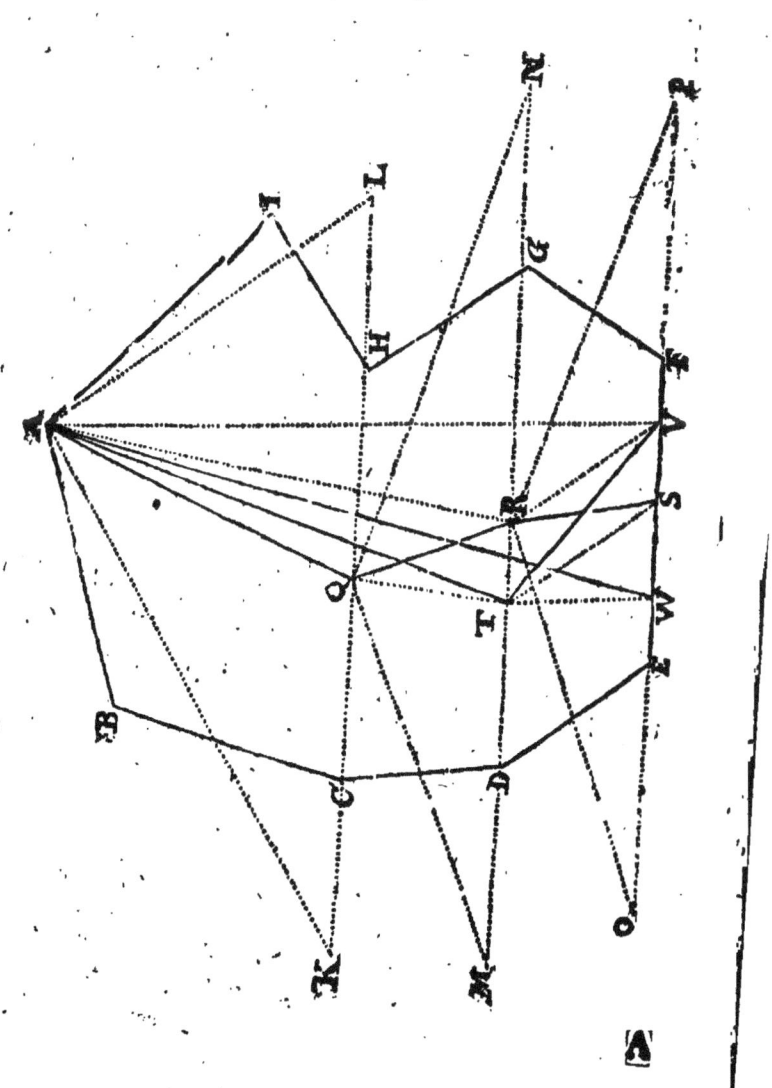

ABCDEFGHI is a Field to be divided between two Men in equal Halfs, by a streight Line proceeding from A.

First, Consider how to divide the Field into Five-sided Figures and Trapezia's, that you may the better reduce it into Triangles: As by drawing the Line K L, you cut off the Five sided Figure ABCHI, which reduce into the Triangle A K L, and measuring half the Base thereof, which will fall at Q, draw the Line Q A.

Secondly, Draw the Line M N, and from the Point Q reduce the Trapezia C D G H into the Triangle M N Q; which again divide into Halfs, and draw the Line Q R.

Thirdly, From the Point R, reduce the Trapezia D E F G into the Triangle R O P; and taking half the Base thereof, draw the Line R S; and then have you divided this irregular Figure into two Equal Parts by the three Lines A Q, Q R, R S.

Fourthly, Draw the Line A R, also Q T parallel thereto. Draw also A T, and then have you turned two of the Lines into one.

Fifthly, From T draw the Line T S; and parallle thereto the Line R V. Draw also T V. Then is your Figure divided into two Equal Parts, by the two Lines A T and T V,

Lastly, Draw the Line A V, and parallel thereto T W. Draw also A W, which will cut the Figure into two Equal Parts by a streight Line as was required.

You may if you please, divide such a Figure all into Triangles; and then divide each Triangle from the Point where the division of the last fell, and then will your Figure be divided by a crooked Line, which you may bring into a streight one, as above,

This

This above is a good way of Dividing Lands, but Surveyors seldom take so much pains about it. I shall therefore shew you how commonly they abbreviate their Work, and is indeed.

An easie way of dividing Lands.

Admit the following Figure A B C D E contain 46 Acres, to be divided into Halfs between two Men, by a Line proceeding from A.

Draw first a Line by guess, through the Figure, as the Line A F. Then cast up the Content of either Half, and see what it wants, or what it is more then the true Half should be.

As for *Example*. I cast up the Content of A E G, and find it to be but 15 Acres? whereas the true Half is 23 Acres; 8 Acres being in the part A B C D G, more than A E G. Therefore I make a Triangle containing 8 Acres, and add it to A E G, as the Triangle A G I; then the Line A I parts the Figure into equal Halfs.

But more plainly how to make this Triangle; Measure first the Line A G, which is 23 Chains, 60 Links. Double the 8 Acres; they make 16; to which add five Cyphers to turn them into Chains and Links, and then they make 1600000; which divide by AG 2360, the Quotient is 6 Chains, 77 Links; for the Perpendicular H I, take from your Scale 6 Chains 77 Links, and set it so from the Base A G F, that the end of the Perpendicular may just touch the Line E D, which will be at I. Then draw the Line A I which makes the Triangle A C I

just

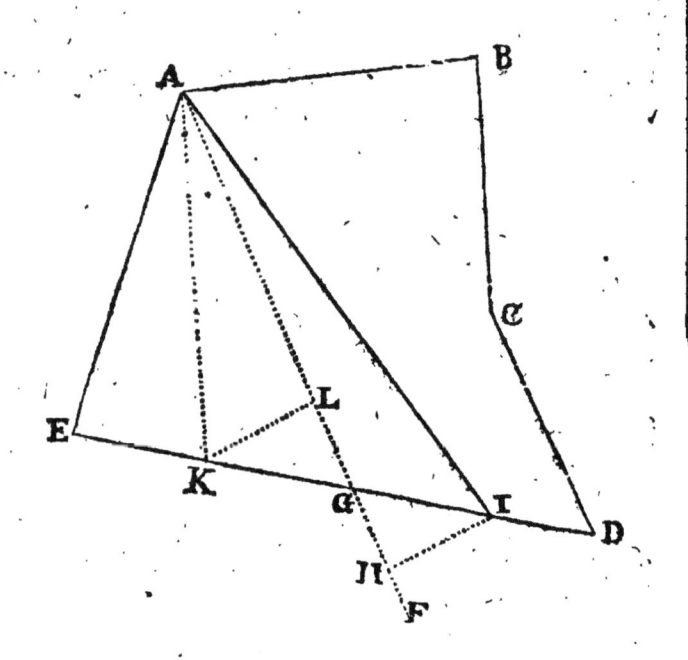

juſt 8 Acres, and divides the whole Figure, as deſired.

If it had been required to have ſet off the Perpendicular the other way, you muſt ſtill have made the end of it but juſt touch the Line E D, as L K does: For the Triangle A K G is equal to the Triangle A G I, each 8 Acres.

And thus you may divide any piece of Land of never ſo many Sides and Angles, according to any Proportion, by ſtreight Lines through it, with as much certainty, and more eaſe, than the former way.

Mark, you might alſo have drawn the Line A D, and meaſured the Triangle A G D, and afterwards have divided the Baſe G D, according to Proportion,

in

Of Dividing of Lands. 157

in the Point I: which I will make more plain in this following Example.

Suppose the following Field, containing 27 Acres, is to be divided between three Men, each to have Nine Acres, and the Lines of Division to run from a Pond in the Field, so that every one may have the benefit of the Water, without going over one another's Land:

First from the Pond ⊙ draw Lines to every Angle, as ⊙A, ⊙B, ⊙C, ⊙D, ⊙E, and then is the Figure

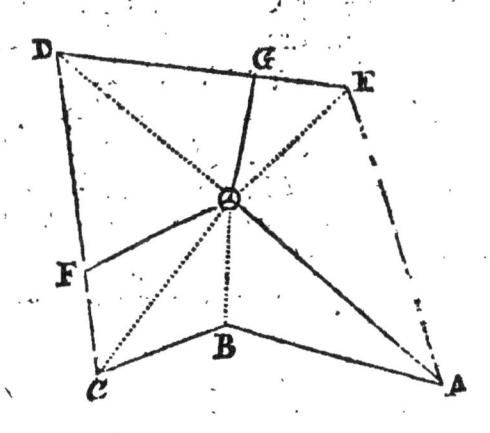

divided into five Triangles, each of which measure and put the Contents down severally; which Contents reduce all into Perches, so will the Triangle.

$$\left.\begin{array}{l} A \odot B \\ B \odot C \\ C \odot D \\ D \odot E \\ E \odot A \end{array}\right\} be \left\{\begin{array}{l} 674 \\ 390 \\ 1238 \\ 911 \\ 1107 \end{array}\right\} Perches.$$

the whole Content being 4320 Perches, or 27 Acres, each Man's Proportion being 1440 Perches.

From

From ⊙ to any Angle draw a Line for the first Division-line, as ⊙ A. Then consider that the first Triangle A ⊙ B is but 674 Perches, and the second B ⊙ C 390, both together but 1064 Perches, less by 376 than 1440, one Man's Portion. You must therefore cut off from the third Triangle C ⊙ D 376 Perches for the first Man's Dividing-line; which thus you may do: The Base D C is 18 Chains; the Content of the Triangle 1238 Perches: Say then, If 1238 Perches give Base 18 Chains, 00 Links, What shall 376 Perches give? Answer, 5 Chains, 45 Links; which set from C to F, and drawing the Line ⊙ F, you have the first Man's part, viz. A ⊙ F.

Secondly, See what remains of the Triangle C ⊙ D 376 being taken out, and you will find it to be 861 Perches, which is less by 578 than 1440. Therefore from the Triangle D ⊙ E cut off 578 Perches, and the point of Division will fall in G. Draw the Line ⊙ G, which with ⊙ A and ⊙ F divides the Figure into three Equal Parts.

How to Divide a Circle according to any Proportion, by a Line Concentrick with the first.

All Circles are in Proportion to one another as the Squares of their Diameters; therefore if you divide the Square of the Diameter or Semi-diameter, and extract the Root, you will have your Desire.

EXAMPLE.

Let A B C D be a Circle to be equally divided between two Men.

The

Of Dividing Land.

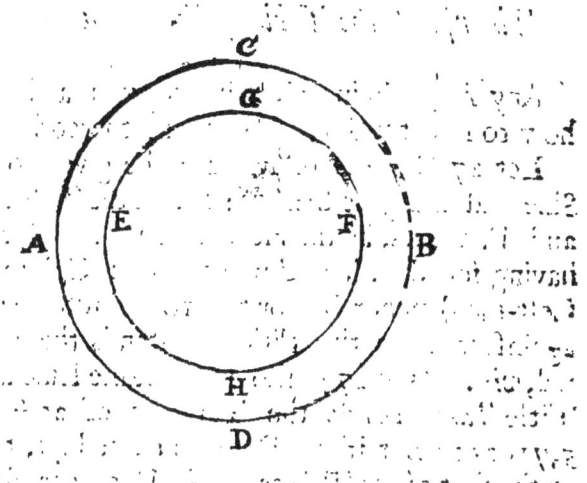

The Diameter thereof is 2 Chains.
The Semidiameter 1 Chain, or 100 Links.
The Square thereof 10000.
Half the Square. 5000.
The Root of the Half 71 Links, which take from your Scale, and upon the same Centre draw the Circle G E H F, which divides the Circle A B C D into two Equal Parts.

CHAP. XII.

Trigonometry: Or the Mensuration of Right Lined Triangles.

The Use of the Table of Logarithm Numbers, I have shewed you in Chap. I. concerning the Extraction of the Square Root. Here follows

The

The use of the Table of Sines and Tangents.

Any Angle being given in Degrees and Minutes, how to find the Sine or Tangent thereof.

Let 27 Degrees 10 Minutes be given to find the Sine and Tangent thereof; first in the Table of Sines and Tangents, at the Head thereof seek for 27, and having found it, look down the first Column on the Left-hand under M for the 10 Minutes, and right against it under the Title *Sin.* stands the Sine required, *viz.* 9,659517; also in the same Line under the Title *Tang* stands the Tangent of 27° 10, *viz.* 9,710282: But if the Degrees exceed 45, then look at the foot of the Tables for the Degrees and upon the Right-hand Column for the Minutes; and right against it you will find the Sine and Tangent above the Title *Sine Tan.* Thus the Sine of 64° Degrees 50 Minutes is 9,956684, the Tangent thereof is 10,328037.

How to find the Co-sine or Sine Complement; the Cotangent, or Tangent Complement of any given Degrees and Minutes.

The Co-sine or Cotangent, is nothing more but the Sine and Tangent of the remaining Degrees and Minutes after substraction from 90 thus, take 25 Degrees 10 Minutes from 90 Degrees, 00 Minutes, there will remain 64 Degrees 50 Minutes, the Sine of which, is as before 9,956684, and that is the Sine Complement of 25 Degrees 10 Minutes.

But

Trigonometry.

But the more ready way to find the Co-sine, or Co-tangent of any number of Degrees given, is to look for the Degrees and Minutes as before taught, for Sines and Tangents, and right against it under Titles Co-sine and Co-Tangent; or above, if the Degrees exceed 45, you will find the Co-sine, or Co-tangent required: Thus the Co-sine of 30 Degrees 15 Minutes, is 9.936431; the Co-tangent of 58 Degrees, 10 Minutes is 9,792974.

Any Sine or Tangent, Co-sine or Co-tangent being given, to find the Degrees and Minutes belonging thereto.

This is only the converse of the former, for you must seek in the Tables for the Sine, &c. given or the nighest that can be found thereto, and right against it you will find the *Minutes*, and *Degrees* over-head. Let the Sine 8,742259 be given, right against it stands 3 *Degrees* 10 *Minutes*.

Remember that Multiplication is performed with these Logarithm Tables by Addition, and Division by Subtraction. If I were to Multiply 5 by 4, first I look for the Logarithm of 5, which is 0,698970 The Logarithm of 4 is   0,602060

Added together, they make   1,301030

which 1301030 I seek for in the Logarithm Tables, and right against, under Title *Num·* stands 20, the Product of 5 multiplied by 4.

If I were to divide 20 by 5, first I look for the Logarithm of 20, which as above, is 1,301030
The Logarithm of 5 is 0,698970

After Subſtraction remains 0,602060

and the Number anſwering to that Logarithm, you will find to be 4.

And thus by Addition and Subſtraction the Rule of Three is performed with the Logarithms, *viz.* By adding the two laſt together, and out of their Product ſubſtracting the firſt.

EXAMPLE.

If 15 give 32, what ſhall 45 give?
The Logarithm of 15 is 1,176091

The Logarithm of 45 is 1,653212
The Logarithm of 32 is 1,505150

The two laſt added together, make 3,158362

Out of which I ſubtract the firſt, and there remains } 1,982271

Againſt which 1,982271, I find the Number 96. I anſwer therefore, If 15 gives 32, 45 ſhall give 96.

This you muſt obſerve to do in the following Caſes of Triangles, always to add the ſecond and third numbers together, and from their Product to Subſtract the firſt, the remainder will be the Logarithm Number, Sine, or Tangent, of your required Line or Angle.

Certain

Certain Theorems for the better understanding Right-Lined Triangles.

1. A Right-lined Triangle is a Figure Comprehended within three streight Lines.

2. It is either Right-Angled, as A, having one Right Angle, which contains just 90 Degrees *viz.* That at *b*; or else Oblique as B, which consists of three Acute Angles, neither of them so great as 90 Degrees; or which consists of two Acute Angles and one Obtuse *viz.* at that D.

3. All the three Angles of any Triangle are equal to two Right Angles, or 180 Degrees; so that one Angle being known, the other two together are known also; or two being known, the third is also known by Substracting the two known Angles out of 180 Degrees, the remainder is the third Angle.

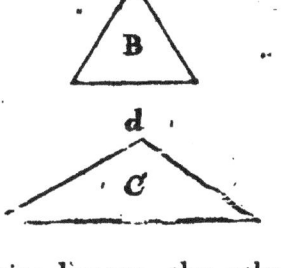

To known well what the Quantity of the Angle is, take this following Domonstration.

Let A B C D be a Circle, whose Circumference is divided (as all Circles you must esteem so to be) into 360 Equal Parts, which are called Degrees and each of those Degrees into 60 Equal Parts more, which are called Minutes: Now a Right-angled Triangle is that which cuts off one fourth part of this Circle, *viz.* 90 Degrees, as you see the Triangle EFG to do.

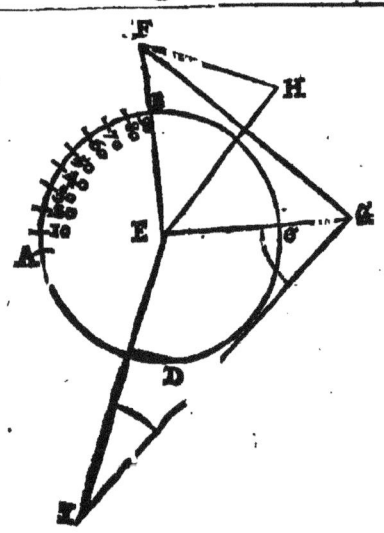

An Angle that cuts off less than 90 Degrees, is called an Acute Angle, as HEF,

GEI is an obtuse Angle, because the two Lines that proceed from E, take in between them more than a quarter of the Circle.

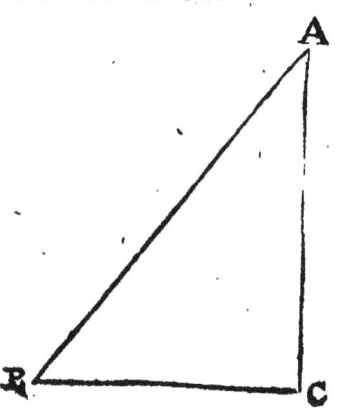

from

Trigonometry.

5. Every Triangle hath 6 Parts, *viz* three Sides and three Angles; the Sides are sometimes called Legs, but most commonly in Right-Angled Triangles the bottom Line, as B C is called the Base A C the Perpendicular, and the longest Line A B is called the Hypothenuse. The Sides are in all in proportion to the Sines of their opposite Angles; so that any three parts of the six being known, the rest may easily be found.

6. When an Angle exceeds 90 Degrees, substract it out of 180, and work by the remainder.

CASE 1.

In Right-Angled Triangles, the Base being given, and the Acute Angle at the Base; how to find the Hypothenuse and the Perpendicular.

In the Right-Angled Triangle A B C, there is given the Base A B, which is 26 Equal Parts, as Perches, or the like; the Angle at A is also given, which is 30 Degrees: Now to find the length of the Hypothenuse A C say thus,

As the Sine Complement of the Angle at A
   is to the Logarithm of the Base 26,
So is Radius or the Sine of 90°
   to the Logarithm of the Hypothenuse A C 30,

The Sine Complement of 30 Degrees is	9,937531
The Logarithm of 26 is	1,414973
The Radius, or Sine of 90°	10,000000
The two laſt added together	11,414973
Remains, after ſubſtracting the firſt Numb.	1,477442

Which if you look for in your Logarithm Tables, you will find the neareſt Number anſwering thereto to be 30. and ſo long is the Hypothenuſe required.

Note in your Tables, when you cannot find exactly the Logarithm you look for, you muſt take the neareſt thereto, as in this Example I find 1,477121 to be the neareſt to 1477442. Mark alſo, that whereas I ſay, as the Sine Complement of the Angle at A, &c. you may as well ſay, as the Sine of the Angle at C is to the Log. &c. For the Angle at A being given in a Right-angled Triangle, you cannot be ignorant of the Angle at G If you mind the Rule above, that all the three Angles, of a Triangle are equal to two right Angles, or 180 Degrees; for if you take the the Right-angled at B 90° and that at A 30° both known, and ſubſtract them out 180°, there remains only 60° for the Angle at C. But in purſuance of our Queſtion.

How to find the Perpendicular

As Sine the of the Angle A C B 60°
  is to the Log of the Baſe 26 A B;
So the Sine of the Angle CAB 30°
  to the Log. of the Perpenicular C B 15.

Note

Trigonometry. 167

Note, when I put three Letters to expreſs an Angle, the middlemoſt Letter denotes the Angular-point,

The Sine 60 *deg*. is   9,937531
The Log. of the Baſe 26 A B, is   1,414973
The Sine of 30 *deg*. is   9,698970

         The two laſt added   11,113943

From which ſubſtract the firſt, and remains 1,176412
The neareſt Number anſwering to which is 15, which is the length of the Perpendicular-line C B.

Or otherwiſe; the Hypothenuſe being firſt found, viz. AC 30 you may find the Perpendicular thus

As the Sine of the Right-Ang. CBA or Rad. 10,000000
    is to the Log. of the Hypoth. A C 30   1,477121
So is the Sine of the angle CAB 30 *deg*.   9,698970

to the Log. of the Perpendicular 15   11,176091

CASE ii

The Perpendicular and Angle ACB being given to find the Baſe and Hypothenuſe

Let the Perpendicular be C B 15, as before the angle ACB 60 *deg*. to find the Baſe, work thus:

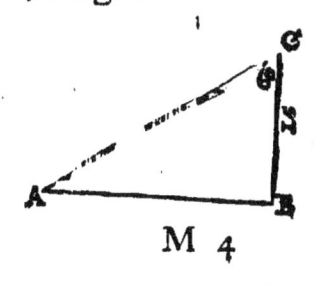

M 4   A

Trigonometry.

As the Co-sine of the Angle A C B
   is to the Logarithm of the Perpendicular B C 15;
So is the Sine of the Angle A C B
   to the Logarithm of the Base A B 26.

The Co-sine of the Angle A C B 60°, is	9,698970
The Logarithm of C B 15 is	1,176091
The Sine of the Angle A C B 60, is	9,937531
	11,113622
The nearest Log. answering to 26, is	1,414652

For the Hypothenuse.

As the Sine complement of the Angle A C B 60°
   is to the Log. of the Perpendicular C B 15
So is the Sine of the Angle A B C, or Radius 90°
   to the Log. of the Hypothenuse 30°

The Co-sine of the Angle A C B, is	9,698970
The Log. of the Perpend, C B 15, is	1,176091
The Radius	10,000000
The Log. of the Hypothenuse 30	1,477121

Or otherwise thus; the Base being first found, to find the Hypothenuse.

As the Sine of the Angle A C B 60,	9,937531
is to the Log. of the Base 26	1,414973
So is Radius	10,000000
to the Log. of the Hypothenuse (30)	1,487442

Case

Trigonometry. 169

CASE iii

The Hypothenuse, and either of the Acute Angles given, to find the Base and Perpendicular.

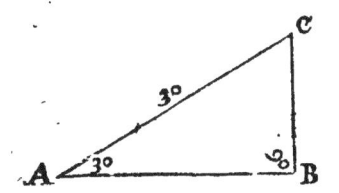

Let the Hypothenuse be A C 30;
The Angle C A B 30

To find the Base AB, work thus:

As the Sine of the Right-Angle } 10,000000
   C B A 90°, or Radius
is to the Log. of the Hypoth. A C 30   1,477121
So is the Co-sine of the angle C A B 30   9,937531

to the Log. of the Base A B (26)   11,414652

To find the Perpendicular B C, work thus.

As the Sine of the Right-Angle } 10,000000
   C B A 90°, or Radius
is to the Log. of the Hypoth A C 30   1,477121
So is the Sine of the Angle C A B 30   9,698970

to the Log. of the Perpend. (15)   11,176091
                                       Or

Or otherwiſe; the Baſe being found, to find the Perpendicular thus.

As the Co-ſine of the Angle C A B 30. 9,937531
  is to the Log. of the Baſe AB 26   1,414973
So is the Sine of the Angle C A B (30.) 9,698970
                                      ─────────
                                      11,113943

to the neareſt Log. of the Perpend. (15) 1,176412

CASE iv.

The Hypothenuſe and Baſe being given, to find the two Acute Angles, viz. A C B, and C A B.

Let A C, the Hypothenuſe, be 30.
A B the Baſe 26. and the Angle A C B required.

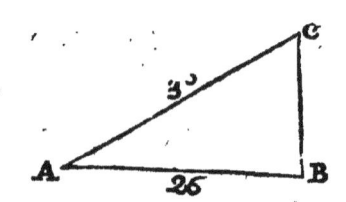

As the Logarithm of the Hypothenuſe A C 30
  is to Radius, or the Sine of the Angle C B A 90
So is the Logarithm of the Baſe A B 26
  to the Sine of the Angle A C B 60.

The

Trigonometry. 171

The Opération.

The Logar. of the Hypothenuse AC 30 is 1,477121
The Radius.   10,000000
The Logarithm of the Base A B 26   1,414973

The Sine of ACB, the Angle required, 60° 9,937852

For the Angle CAB, work thus.

As the Logar, of the Hypothenuse AC 30   1,477121
  is to the Radius 90   10,000000
So is the Logarithm of the Base A B 26   1,414973

to the Co-sine of the Angle required 30 9,937852

CASE V.

The Hypothenuse and Perpendicular being given, to find the Angles and Base.

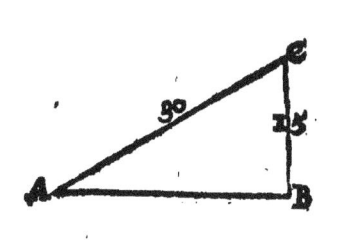

The Hypothenusal is 30
The Perpendicular 15
A B C a Right Angle.

Now.

Now to find the Angle at A, work thus.

As the Logar of the Hypothenuse A C 30 1,477121
   to the Raduis.   10,000000
So is the Logar. of the Perpendicular 15 CB 1,176091

To the Sine of the Angle at A 30   9,698970

To find the Angle at C, work thus.

As the Logarithm of the Hyponthenuse A C 30
Is to Radius 90
So is the Logarithm of Perpendicular B C 15
To the Co-sine of the Angle · which is the Angle C 60
   Lastly, to find the Base, work as you were taught in Case 2,
Here Note, that any two Sides of a Right-Angled Triangle being given, the third Side may be found by Extraction of the Square Root.

EXAMPLE:

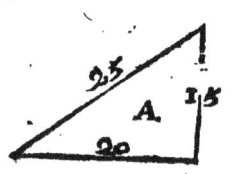

In the Right-angled Triangle A, let the given Base be 20, the Perpendicular 15, and the Hypothenuse required.
Square the Base 20, or multiply it by itself, and it makes 400; Square also the Perpendicular 15, and it makes 225, add the two Squares together, and they make 625, from which Summ extract the Square Root, which Root is the length

length of the Hypothenuſal, *viz.* 25;
but if the Hypothenſual, and either
of the other Sides be given to find
the third, you muſt Subſtract the
Leſſer Square out of the Greater, and the Root of the remainder is the Side required: As for Example, the Hypothenuſe 25 is given, and the Baſe 20, to find the Perpendicular multiply the Hypothenuſe in itſelf, and it makes.
Multiply the Baſe in itſelf, and it makes

$$ 20 \\ 625\ (25 \\ 45$$

$$625 \\ 400$$

which 400 ſubſtract from 625, there remains 225

the Root of which is 15, the Perpendicular required.

CASE vi.

Of Oblique-Angled Plain Triangles.

Two Sides of an Oblique-Triangles being given, and an Angle oppoſite to either of the Sides, how to find the other two Angles and the third Side.

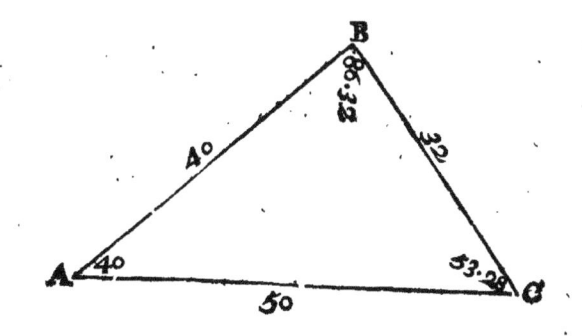

In the Triangle A B C there is given the Side A B 40, the Side B C 32;
the Angle at A 40 Degrees,
and the Angle at C is required.

Note, that in Oblique-Triangles, the same Rule holds good as in Right-ang'ed Triangles; *viz.* That the Sides are in such proportion one to another, as the Sines of their opposite Angles.

As the Logarithm of the Side B C 32   1,505150
  is to the Sine of the Angle A 40   9,808067
So is the Logarithm of the side A B 40   1,602060
                                            11,410127

to the Sine of the Angle at C 53o : 28'   9.904977

To find the Angle at B

Add the two known Angles together, *viz.* that at A 40, and that at C. 53.28, and they make 93 Degrees 28 Minutes; which substracted from 180 Degrees, leaves 86 Degrees 32 Minutes, which is the Angle at B.

Lastly to find the line A C, say,

As the Sine of the Angle A 40   9,808067
  is to the Logarithm of the Side B C 32   1,505150
So is the Sine of the Angle 86o : 32   9,999204
                                         11,504354

to the Log. of the Side A C required 50 1,696287

Note

Trigonometry.

Note, Though the nearest whole Number answering to the Logarithm 1696287 be 50; yet if you go to Fractions, the length of the Line A C is but 49 $\frac{69}{80}$.

CASE vii.

Two Angles being given, and a Side opposite to one of them, to find the other opposite Side.

In the foregoing Triangle there is given the Angle A 40 Degrees, the Angle C 53 Degrees 28 Minutes; also the Side A B 40: To find the Side B C work thus.

As the Sine of the Angle C 53 : 28   9,904992
   is to the Logarithm of the Side A B 40   1,602060
So is the Sine of the Angle A 40   9,808067
                                     11,410127

To the Log of the Side B C, nearest 32   1,505135

CASE viii.

Two Sides of a Triangle being given, with the Angle contained by them, to find either of the other Angles.

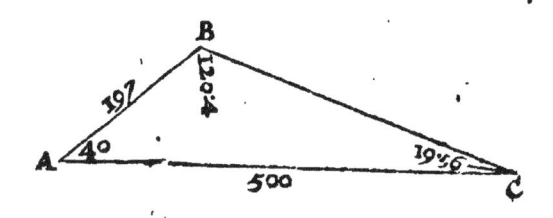

In

In the Triangle A B C
  there is given the Side A B 197
The Side AC 500
The Angle at A 40 Degrees;
Now to find either of the other Angles work thus.
As the Log. of the Sum of the 2 Side 697 2,843233
  is to the Log. of there Difference 303 2,481443
So is the Tang. of the half Sum of the
  two Opposite Angles 70 Degrees } 10,438934

                                        12,920377

  to the Tangent of 50 Degrees 4 Min. 10,077144

which 50° 4' added to the half summ of the two unknown Angles, *viz.* 70° makes 120° 4', which is the Quantity of the Angle at B, also taken from 70, leaves 19 *deg.* 56', which is the Angle at C.

CASE ix.

Three Sides of an Oblique Triangle being given, to find the Angles.

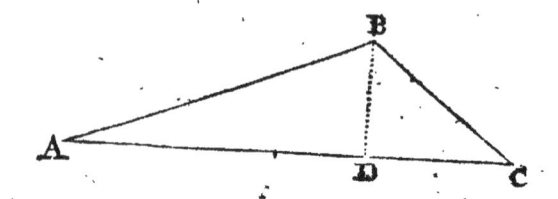

You must first Divide your Oblique Triangle into two Right-angled Triangles thus,

In

Trigonometry.

In the Triangle A B C
The Side A C is   50

The Side A B   36
The Side B C   20

The Summ of the two Lesser Sides   56

The Differance of the two Lesser Sides   16

As the Log. of the greatest Side A C 50   1,698970
  is to the Logar. of the Summ of the two Lesser Sides 56   1,748188
So is the Differ. of the two Lesser Sides 16   1,204120

                                      2,952308

to the Log. of a fourth Number 18   1,253338

Substract this 18 out of the greatest Side A C 50, and there remains 32, the half of which, *viz* 16, is the Base of the Lesser Right-Angled Triangle and the remainder of the Line A C, *viz.* A D 34 is the Base of the Greater Right-Angled Triangle, into which this Oblique Triangle is divided,

And now of either Right-Angled Triangle BDC, or BDA, you have the Base and Hypothenuse given to find the Angles, which you must do as you were before taught, Case iv.

Note that you may better and easier find the 4th Number, for dividing an Oblique-angled Triangle into two Right-Angled Triangles by Vulgar Arithmetik, then by the Tables of Logarithms thus.

                 N   Square

178   *Trigonometry.*

Square the three given Sides, add the two greater Squares together; and from that Summ Subſtract the Leſſer; half the remainder divide by the greater Side; the Quotient will be the Baſe of the Greater Right-Angled Triangle.

EXAMPLE.

In the fore-going Triangle, the Square of the greateſt Side A C 50, is   2500
The Square of the Side A B 36, is   1296

Added together, make   3796

From which ſubſtract the Square of the leaſt Side.   400

Remains   3396

The Half   1698

Which 1698 divide by 50 the longeſt Side; the Quotient is 33 $\frac{48}{50}$, the Baſe of the greater Right-Angled Triangle, *viz.* A D; and that being ſubſtracted out of 50, leaves 16,$\frac{2}{50}$ for the Baſe of the ſmaller Right-Angled Triangle, *viz.* D C.

CASE

CASE I.

The three Sides of an Oblique Triangle being given, how to find the Superficial Content without knowing the Perpendicular.

From half the Summ of the Three Sides, substract each particular Side. Add the Logarithms of the three Differences, also the Log. of half the Sum of the three Sides together. Half the Total is the Log. of the Content required.

In the foregoing Triangle, the Sides are 50, 36, 20, their Sum is 106: The half Sum 53.

The differences between the half Sum and each particular Side, are

	3	Log. 0.477121
	17	1.230449
	33	1.518514
The half Sum | 53 | 1.724276 |

Total added 4.950360

The Half 2.475186

The Number answering to that Log. is 298 which is the Content of the Triangle required.

By Vulgar Arithmetick, thus.

Multiply the First Difference by the Second; that Product by the third; that Product by the HalfSum. Lastly, Extract the Square-Root, and you have the Super-

Superfical Content. So 3 multiplied by 17. makes 51; which multiplied by 33, makes 1683. that multiplied by 53, the half Sum makes 89199. the Square-Root of which is 298, the Content required.

CHAP. XII

Of Heights and Distances.

How to take Height of a Tower, Steeple, Tree, or any such thing.

LEt A B be a Tower whose Height you would know.

First, at any convenient distance, as at C, place your Semi-circle, or what other Instrument you judge most fit for the taking an Angle of Altitude, as a large Quadrant, or the like, and there observe the Angle A C B. But to be more plain, place your Semi-circle at C; and having turn'd it down by a Plumb, make it to stand Horizontal, which it does when a Plummet-line fixt to the Centre falls just upon 90 *deg.* (in some Semi-circles there is a Line on the Back-side of the Brass Limb on purpose for the setting it Horizontal.) Then first screwing the Instrument fast) move the Index up and down, till through the Sights you espy the top of the Tower at A. See then what Deg. upon the Limb are cut by the Index,
which

Heights and Distances.

which let be 58, so much is your Angle of Altitude. Measure next the distance between your Instrument and the foot of the Tower, viz. the Line C D, which let be 25 Yards: Then have you all the Angles given, (admitting the Angle of the Tower makes with the Ground, viz. d to be the Right-Angle) and the Base to find the Perpendicular A B; which you may do, as you were taught in *Case* I. of *Trigonometry*: For if you take 58 from 90, there remains 32 for the Angle at A. Then say,

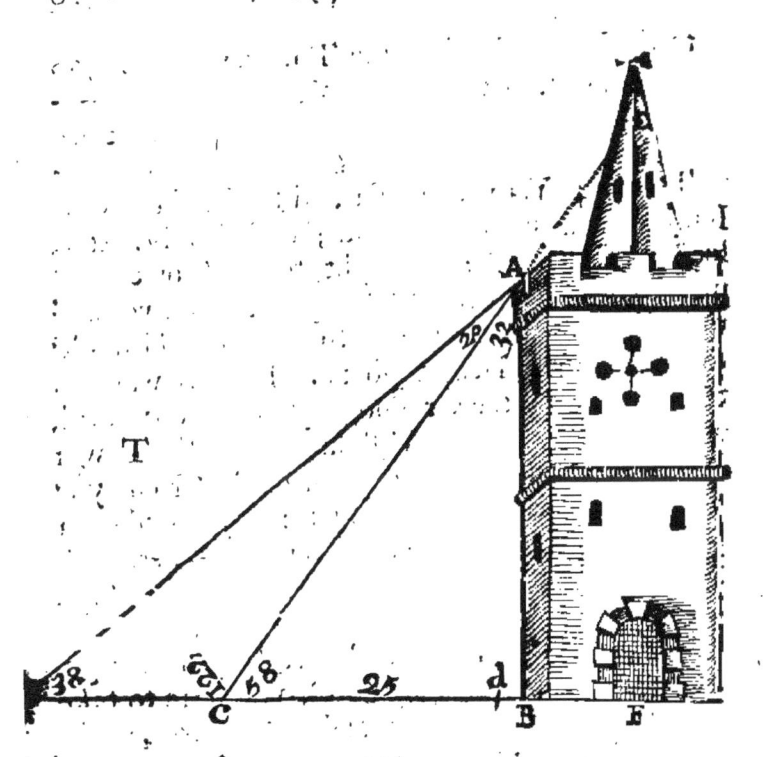

As the Sine of the Angle A 32	9,724210
is to the Log. of the Base C D 25	1,397940
So is the Sine of the Angle C 58	9,928420
to the Log. Heighth of the Tower, AB, or rather AD, 40 Yards.	11,326360
	1,602150

To this 40 Yards you must add the height of your Instrument from the Ground; or which is better, look through your Fixed-Sight to the Tower, and mark where your Sight falls upon the Tower and measure from that place to the ground, which add to the former Height found. In this way of taking Heighths, the Ground ought to be very level, or you may make great Mistakes. Also the Tower or Tree should stand perpendicular: Or else you must measure to such a place, where a Perpendicular would fall if let down; as A B is not a Perpendicular; but A d, therefore measure the Distance C d, for your Base.

This you may plainly understand by the foregoing Figure; for if standing at C, you were to take the Height of the Tower and Steeple to E: The Angle ECB is the same as the Angle at ACB; and if you measure only CB or CD, you will make the Heighth FE the same as DA; which by the Figure you plainly perceive to be a great Error: Therefore to take the Heighth FE you should measure from C to F.

Heigths and Distances.

How to take the Height of a Tower, &c. when you cannot come nigh the Foot thereof.

In the foregoing Figure, let A B be the Tower, and suppose C B to be a Moat, or some other hindrance, that you cannot come nigher than C to take the Height. Therefore at C place your Instrument, and take (as before) the Angle ACB 58 *deg.* Then go backwards any convenient distance, as to G, there also take the Angle AGB 38 *deg.* This done substract 58 from 180, so have you 122 *deg.* the Angle ACG. Then 122 and 38 being taken from 180, remain 20 for the Angle GAC. The Distance GC measured, is 26. Now by *Trigonometry*, say,

As the Sine of the Angle A 20	9534052
is to the Log. of the Distance G C 26	1414973
So is the Sine of Angle G 38	9789343
	11204315
to the Log. of the Line A C 47	1,670263

Again,

As Radius the Right-Angle B	10,000000
is to the Log. of the Line AC 47	1,672098
So is the Sine of the Angle C 58	9,928420
To the Log. Heigth of the Tower 40 Yards	11,600518

But still, as I told you before, the Ground is understood to be level. However, if it be not, I will shew you,

along one Side of it, would as well know the Breadth of it, as also make a true Plot thereof, by putting down what remarkable things are seen on the other Side.

Beginning at ☉ 1, the first Station, cause one of your Assistants to go to the next bend of the River, as ☉ 2, and there set up a Mark for you; then see what Angle from the Meridian ☉ 1, ☉ 2 makes, which let be N. 6 Degrees W.; also seeing several Marks on the other Side of the River, taking their Bearings, as the House A, which stands upon the Bank, and is a good Mark, for the Breadth of the River bears N. W. 52 deg. the Wind-mill B up in the Land, bears N. W. 40 deg. The Tree C by the Water-side, bears N. W. 17 deg. All this note down in your Field-Book, and measure the distance ☉ 1, ☉ 2, 18 Chains, 20 Links. After this, coming to ☉ 2, see how the next bend of the River bears from you, viz. ☉ 3; which is N. E. 15 deg. See also how the House A there bears from you, viz. S. W. 20 deg. The Wind-mill S. W. 50 deg. The Tree N. W. 77. Also as you are going forward, if you see any thing more at this second Station, taking the bearing thereof, as a noted House D up in the Land, bears N. W. 28° And a Church E close by the Rivers brink N. W. 4° measure the distance 2, 3, and placing your Instrument at 3, the Church bears from you N. W. 88 deg. The House up in the Land D you connot see for the Church, therefore let it alone for the next Station. But here you may see forward a little Village F, the first House whereof bears from you N. W. 32 deg. Measure the distance 3, 4, and planting your Instrum. in 4, the first House of the Village F bears from you S. W. 32 deg. and the House D, which you could

not

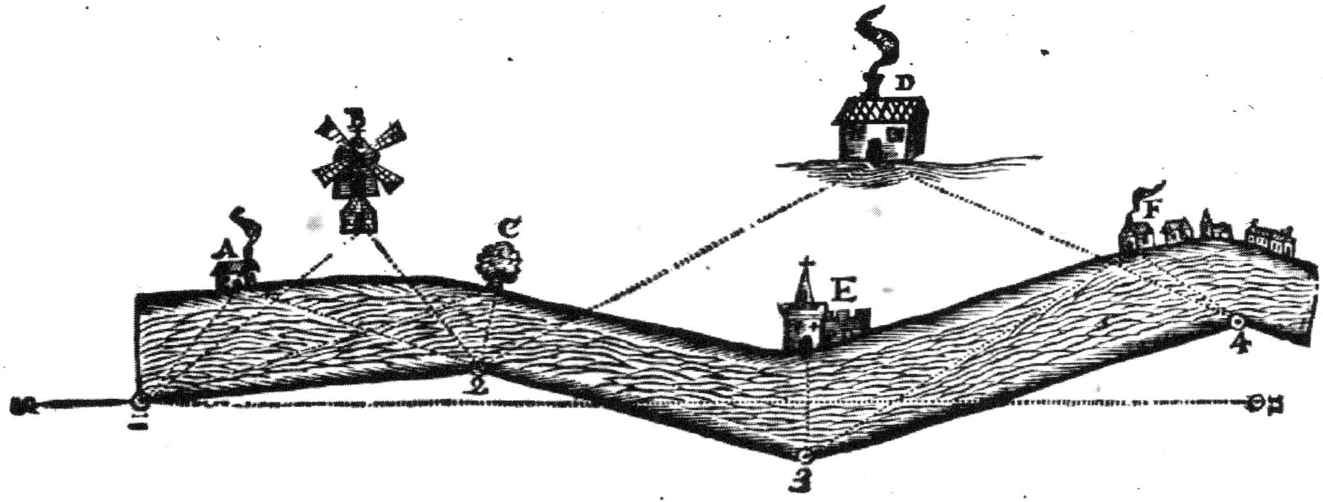
A
B
C
D
E
F
1
2
3
4

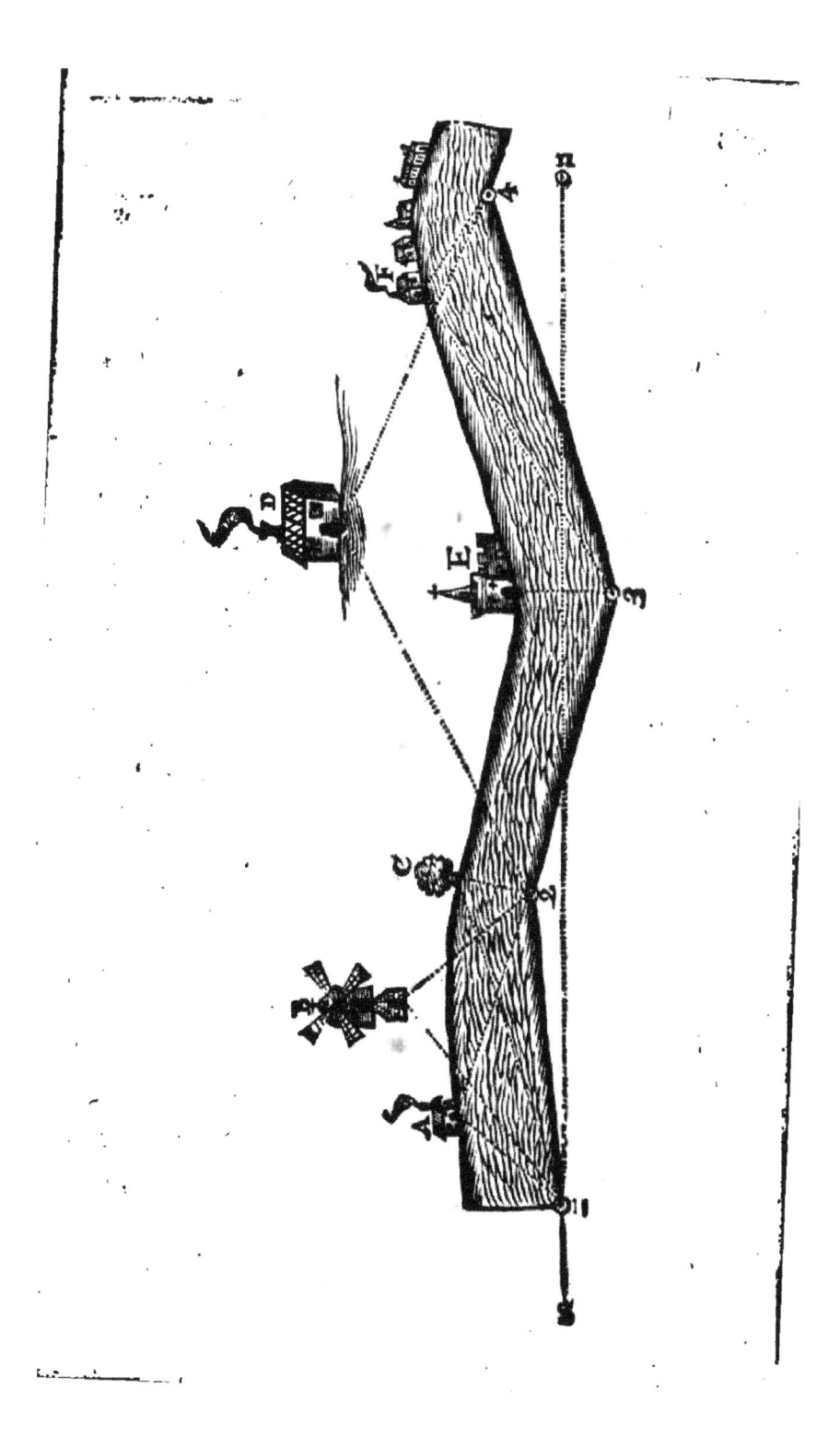

not see at the third Station, S. W. 24° Having put down all these things in your Field Book, it will look thus,

☉ Observation	1 N. W. 60. 18. Ch. 20 Lin.
	A Tree upon the brink of the River bears N. W. 17° 00′
	A Wind-mill up in the Land N. W. 40° 00′
	A House upon the Riv bank N. W. 52° 00′

☉ 2 N. E. 15° 18 Ch. 10 Lin.
{ The Tree   N. W. 77° } These look back to
{ The House   S. W. 26° } the Observation of
{ The Wind-mill S. W 50°   } ☉ 1.
{ A noted House far in up the }
{   Land   N. W. 28° } Forward Observa-
{ A Church upon the River's } tions.
{   bank   N. W. 4° }

☉ 3 N. W. 15, 20 Ch. 50 Lin,
{ The Ch. bears N. W. 88°   } These look back
{ The noted H. cannot be seen } to the Ob. of ☉ 2.
{ The end of a little Village   }
{   N. W. 32 } A forward Observat

☉ 4 ——————————
{ The end of the little Village }
{   S. W. 32   } These respect ☉ 3
{ The House respecting ☉ 2 } and ☉ 2
{   in the Land   S. W. 24° }

To Protract this, draw the Line N S for a Meridian, and laying your Protractor upon it, the Centre thereof to ☉ 1; against N W 6 make a Mark for the Line that goes to ☉ 2: Also against N W 17 make a Mark for the Tree, and against 40 and 52,
for

Heigths and Distances. 489

for the Wind-mill and House. Then from ☉ 1
through these Marks draw the Lines ☉ A, ☉ B, ☉ C, ☉ 2.
  Secondly, Take from your Scale 18 Ch. 26 Lin.
and set it off upon the Line ☉ 2, which will reach
to ☉ 2. There lay again the Centre of your Protractor, the Diameter thereof parallel to the Line NS,
and make Marks, as you see in the Field-Book, against
NE 15. NW 77. SW 20. SW 50. NW 28.
NW 40. and through these Marks draw Lines.
The first Line directs to your third Station; the second Line NW 77. directs you to the Tree C upon
the Rivers bank for that Line cutting the Line ☉ 1
C, shews you by the Intersection where thee Tree
stood, and also the Breadth of the River. Also the
Line SW 20 cuts the Line from the first Station
NW 52, in the place where the House A stands
upon the Bank of the River. If therefore you draw
a Line from A to C, it will represent the farther
Bank of the River. And so you may proceed on
Plotting, according to the Notes in your Field-
Book; and you will not only have a true Plot of
the River, but also know how far the Wind-mill B,
and the House D, stand from the Water-side.

How to take the Horizontal-line of a Hill.

  When you measure a Hill, you must measure the
Superficies therefore, and accordingly cast up the Contents. But when you Plot it down, because you
cannot make a Convex Superficies upon the Paper,
you must only Plot the Horizontal or Base thereof;
which you must shadow over with the resemblance
of a Hill, that other Surveyors, when they apply
your

your Scale therto, may not say you were Miſtaken. And you may find this Horizontal or Baſe-line after the ſame manner as you have been taught to take Heighths.

For ſuppoſe ABCD a Hill, whoſe Baſe you would know. Plant your Semi-circle at A, and cauſe a Mark to be ſet up at B, ſo high above the top of the

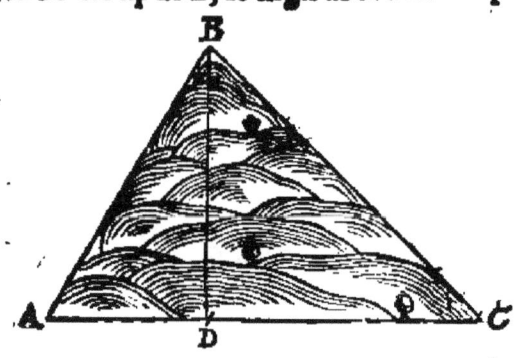

Hill, as the Inſtrument ſtands from the Ground at A; and making your Inſtrument Horizontal, take the Angle BAD 58 *deg*. Meaſure the Diſtance A B 16 Chains, 80 Links. Then ſay,

As Radius	10000000
is to the Line A B 16 Ch. 80 Lin.	3225309
So is the Sine Complement of A 58°	9724210
to part of the Baſe AD 8 Ch. 90 Lin.	12,949519

But if you have occaſion to meaſure the whole Hill, plant again your Inſtrument at B, and take the Angle C BD, which let be 46 *deg*. Meaſure alſo the Diſtance B C 21 Ch. Then ſay,

As

Heights and Distances.   191

As Radius	10000000
is to the Line BC 21 Ch. 00 Lin. (Log.)	1322219
So is the Sine of the Angle CBD 46	9856934
to the part of the Base DC 15 Ch. 11 Lin.	11,179153

Which 15. 11. added to 8. 90, makes 24 Chains, 1 Link, for the whole Base AG; which is to be plotted, and not AB and BC; although they are to be measured to find the Content of the Land.

I mentioned this way, for your better understanding how to take the Base of part of a Hill; for many times your Survey ends upon the Side of a Hill. But if you find you are to take in the whole Hill, you need not take altogether so much pains as by the former way. As thus: Take, as before the Angle A 58 deg. Measure also AB. Then at B take the whole Ang. ABC 78 deg. Substract these two from 180 deg. remains 44 for the Angle at C. Then lay,

As the Sine of the Angle C 44
   is to the Log. of the Side AB;
So is the Sine of the Angle ABC
   to the Log. of the Base AC.

How to take the Shoals of a River's Mouth, and Plot the same.

Measure first the Sea coast on both Sides of the River's Mouth, as far as you think you shall have occasion to make use thereof; and make a fair Draught thereof, putting down every remarkable thing in its true Situation, as Trees, Houses Towns Wind-mill, &c. Then going out in a Boat to such
                                                  Sands

Sands or Rocks as make the Entrance difficult, at every confiderable bend of the Sands, take with a Sea-Compafs the bearing thereof to two known Marks upon the Shore, and having fo gone round all the Sands and Rocks, you may eafily upon the Plot before taken, draw Lines which fhall interfect each other at every confiderable Point of the Sands, whereby you may truly prick out the Sands, and give good Directions either for laying Bouyes, or making Marks upon the Shore for the Direction of Shipping.

EXAMPLE.

Suppofe the following Figure to be a peice of fome Sea-Coaft. Firft I make a fair Draught of it, with the Mouth of the River as far up as there is occafion putting down every remarkable thing, as you fee here, all but the Rocks and Sands excepted, which I am now going to fhew you how to take. Go in a Boat down the River, till you find the beginning of the firft Sand A, as at *a*, and there take a Sight to the Red-Houfe, which let be S. W. 86 *deg*. alfo to the Tree, which is S. E. 6 *deg*. To Plot which draw Lines quite contrary to your obfervations; as from the Red-Houfe, draw a Line N. E. 86, and from the Tree a Line N. W. 6 *deg*. which two Lines will interfect each other in the Point *a*, which fhews you the beginning of the Sand A. Row along the fame Sand, founding as you go, till you find it have a confiderable bending, and there take again two obfervations, as before, and Protract them too, when you come a-fhore, in like manner. The like do at the bending of every Sand, till either you come round

the

Heigbths and Diftances.   193

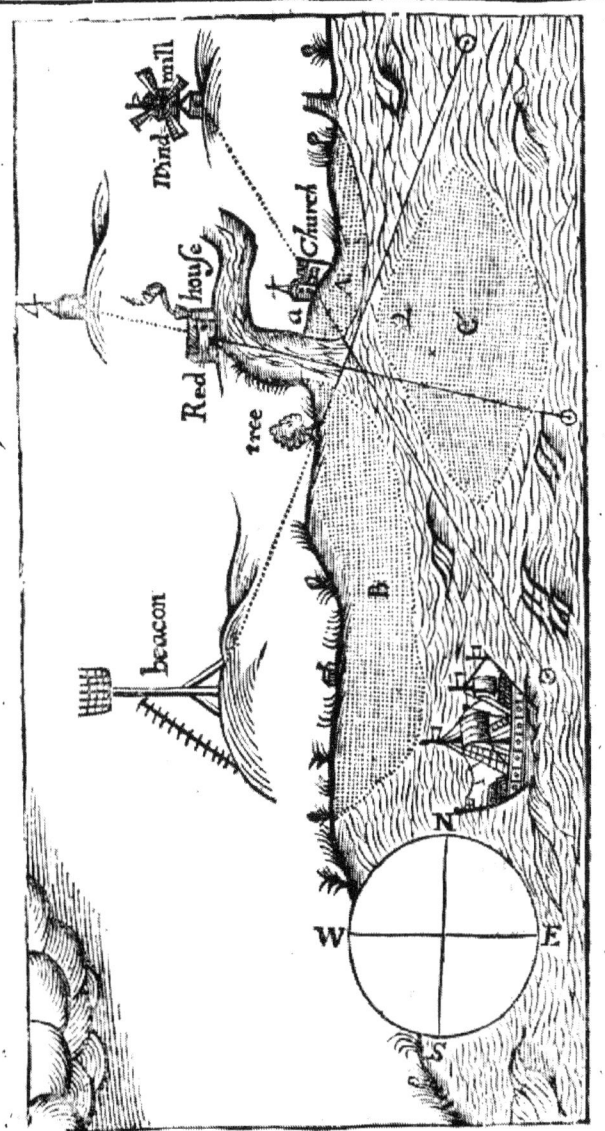

the Sand, or come to the Place where it joins with the Shore.

It would be too tedious for you, and troublefom
fo

for me, to give you all the Obfervations, I having already in this Treatife fo often defcribed the fame thing before; therefore I will mention only one Place of Obfervation more which I Judge fufficient. In the Sand C, I find the bend (2.) and there as I fhould do at all the reft, I take two Obfervations to fuch things on the Shore as are moft confpicuous unto me, *viz.* Firft, to the Beacon, which bears from me S.W. 25 *deg.* Secondly to the Wind-mill, which bears from me N. W. 40 *deg.* Now after I have taken the other Angles or Bends of that Sand, and am come Home, I draw a Line from the Beacon oppofite to my Obfervation S. W. 25 *deg. viz.* N.E. 25 *deg.* Alfo from the Wind-mill I draw a Line S. E. 40 *deg.* Now, where thefe two Lines interfect each other, as they do at 2, I mark for one Point of the Sand C. In like manner as I did this, I obferve and protract every Line of the Sand C, and of all the other Sands, and Rocks, be there never fo many; and fo will you have a fair Map, fitting for Seamen's Ufe.

Now to give Direction for Seamen's coming in here, draw a Line through the middle of the South Channel, which Line will cut both the Church and Wind-mill; fo that if a Ship coming from the Southward, bring the Church and Wind-mill both into one, and keep them fo, fhe may boldly run in, till fhe brings the River's Mouth fair open, and then fail up the River. Likewife coming from the Northward, muft firft bring the Tree and Beacon both into one, and keep them fo till the River's Mouth is fair open. But left they fhould miftake and run upon the ends of the Sands A or B, it would be neceffary that a Mark was fet up behind the Red-Houfe, in a ftreight Line with the middle of the River, as ⚓ Then a Ship coming from the Southward, or Northward,

ward, let her keep her former Marks both in one, till she bring the Red-House and ⚑ both in one; and then keeping them so, run boldly up the River, till all Danger is Past. I have put down this Windmill and Beacon, not as if such good Marks would always happen; but to shew you how to place Marks, or lay Bouys if it be required.

You must mind after you have taken all the Sands, to take the Sounding also, quite cross the Channels, all up and down, and to put them down accordingly; the best time for doing which, is at Low-Water in Spring-Tides.

How to know whether Water may be made to run from a Spring-head to any appointed Place.

For this Work, the Diameter of the Semi-circle is a little too short; however an indifferent shift may be made therewith, but is is better to get a Water-level, such as you may buy at the Instrument-Makers, with which being provided, as also with two Assistants, and each of them with a Staff divided into Feet, Inches, and Parts of an Inch, go to the Spring-head; and causing your first Assistant to stand there with his Staff perpendicular, make the other go in a Right-line towards the Place designed for bringing the Water, any convenient distance, as 100, 150, or 200 Yards, and there let him stand, and hold his Staff perpendicular also. Then set your Instrument nigh the Mid-way between them, making it stand Level, or Horizontal; and look through the Sights thereof to your first Assistant's Staff, he moving a piece of white Paper up and down the Staff, according to the Signs you make to him, till through the Sights you espy the very Edge of the Paper. Then by a Sign make him to understand

that

that you have done with him; and let him write down how many Feet, Inches and Parts, the Paper rested upon. Also going to the other end of your Level, do the same by the second Assistant, and let him write down also what number of Feet, &c. the Paper was from the Ground. This done, let your first Assistant come to the second Assistant's Place, and there let him again stand with his Staff; and let the second Assistant go forward 100 200 Yards, as before; and placing your self and Instrument in the midst between them, take your Observations altogether, as before, and let them put them down in like manner. And so must you do till you come to the Place whereto the Water is to be conveyed. Then examine the Names of both your Assistants, and if the Notes of the second Assistant exceed that of the first, you may be sure the Place is lower then the Spring-head, and that therefore Water may be well conveyed. But if the first's Notes exceed the seconds, you may conclude it impossible, without Engines, or the like.

The first Assistant's Note. Stat. Feet, Inch, Parts.	The second Assistant's Note Stat. Feet, Inch, Parts
⊙1 4 3 5	⊙1 14 5 1
⊙2 2 4 2	⊙2 4 6 3
⊙3 3 5 1	⊙3 9 2 4
20 0 8	28 1 8

Here you may see the second Assistant's Note exceeds the first, 8 Feet, 1 Inch; which is enough to bring the Water with a strong current, and to make it also rise up 6 or 7 Feet in the House, if occasion be; for such as have written of this Matter, allow but 4 Inches and ¼ Fall in a Mile to make the Water run

A TABLE

OF THE

Northing or Southing, Easting or Westing, of every Degree from the Meridian, according to the Number of Chains run upon any DEGREE.

A Table of Northing, or Southing,

Distance	1 Deg. N.S	1 Deg. E.W	Distance	2 Deg. N.S	2 Deg. E.W	Distance	3 Deg. N.S	3 Deg. E.W
1	1.0	.0	1	1.0	.0	1	1.0	.1
2	2.0	.0	2	2.0	.1	2	2.0	.1
3	3.0	.0	3	3.0	.1	3	3.0	.1
4	4.0	.1	4	4.0	.1	4	4.0	.2
5	5.0	.1	5	5.0	.2	5	5.0	.2
6	6.0	.1	6	6.0	.2	6	6.0	.3
7	7.0	.1	7	7.0	.2	7	7.0	.4
8	8.0	.1	8	8.0	.3	8	8.0	.4
9	9.0	.2	9	9.0	.3	9	9.0	.5
10	10.0	.2	10	10.0	.3	10	10.0	.5
20	20.0	.4	20	20.0	.7	20	20.0	1.0
30	30.0	.5	30	30.0	1.0	30	30.0	1.6
40	40.0	.7	40	40.0	1.4	40	40.0	2.1
50	50.0	.9	50	50.0	1.7	50	50.0	2.6
60	60.0	1.1	60	60.0	2.1	60	59.9	3.1
70	70.0	1.2	70	70.0	2.4	70	69.9	3.7
80	80.0	1.4	80	80.0	2.8	80	79.9	4.2
90	90.0	1.6	90	89.9	3.1	90	89.9	4.7
100	100.0	1.8	100	99.9	3.5	100	99.9	5.2

| Dist. | E.W | N.S | Dist. | E.W | N.S | Dist. | E.W | N.S |

| 89 Deg. | 88 Deg. | 87 Deg. |

Easting, or Westing.

Distance	4 Deg. N.S	4 Deg. E.W	Distance	5 Deg. N.S	5 Deg. E.W	Distance	6 Deg. N.S	6 Deg. E.W
1	1.0	.1	1	1.0	.1	1	1.0	.1
2	2.0	.1	2	2.0	.2	2	2.0	.2
3	3.0	.2	3	3.0	.3	3	3.0	.3
4	4.0	.3	4	4.0	.3	4	4.0	.4
5	5.0	.3	5	5.0	.4	5	5.0	.5
6	6.0	.4	6	6.0	.5	6	6.0	.6
7	7.0	.5	7	7.0	.6	7	7.0	.7
8	8.0	.6	8	8.0	.7	8	8.0	.8
9	9.0	.6	9	9.0	.8	9	8.9	.9
10	10.0	.7	10	10.0	.9	10	9.9	1.0
20	20.0	1.4	20	20.0	1.7	20	19.9	2.1
30	29.9	2.1	30	29.9	2.6	30	29.8	3.1
40	39.9	2.8	40	39.8	3.5	40	39.8	4.2
50	49.9	3.5	50	49.8	4.4	50	49.7	5.2
60	59.9	4.2	60	59.8	5.2	60	59.7	6.3
70	69.8	4.9	70	69.7	6.1	70	69.6	7.3
80	79.8	5.6	80	79.7	7.0	80	79.6	8.3
90	89.8	6.3	90	89.7	7.9	90	89.5	9.4
100	99.8	7.0	100	99.6	8.7	100	99.5	10.4

Dist.	E.W	N.S	Dist.	E.W	N.S	Dist.	E.W	N.S
	86 Deg.			85 Deg.			84 Deg.	

A 2

A Table of Northing, or Southing,

Distance	7 Deg. N.S.	7 Deg. E.W.	Distance	8 Deg. N.S.	8 Deg. E.W.	Distance	9 Deg. N.S.	9 Deg. E.W.
1	1.0	.1	1	1.0	.1	1	1.0	.2
2	2.0	.2	2	2.0	.3	2	2.0	.3
3	3.0	.4	3	3.0	.4	3	3.0	.5
4	4.0	.5	4	4.0	.6	4	4.0	.6
5	5.0	.6	5	5.0	.7	5	5.0	.8
6	6.0	.7	6	5.9	.8	6	5.9	.9
7	6.9	.8	7	6.9	1.0	7	6.9	1.1
8	7.9	1.0	8	7.9	1.1	8	7.9	1.3
9	8.9	1.1	9	8.9	1.3	9	8.9	1.4
10	9.9	1.2	10	9.9	1.4	10	9.9	1.6
20	19.9	2.4	20	19.8	2.8	20	19.8	3.1
30	29.8	3.7	30	29.7	4.2	30	29.6	4.7
40	39.7	4.9	40	39.6	5.6	40	39.5	6.3
50	49.6	6.1	50	49.5	7.0	50	49.4	7.8
60	59.6	7.3	60	59.4	8.3	60	59.3	9.4
70	69.5	8.5	70	69.3	9.7	70	69.1	10.9
80	79.4	9.8	80	79.2	11.1	80	79.0	12.5
90	89.3	11.0	90	89.1	12.5	90	88.9	14.1
100	99.3	12.2	100	99.0	13.9	100	98.8	15.6

Dist.	E.W.	N.S.	Dist.	E.W.	N.S.	Dist.	E.W.	N.S.
	83 Deg.			82 Deg.			81 Deg.	

Easting, or Westing.

Distance	10 Deg. N.S.	10 Deg. E.W	Distance	11 Deg. N.S.	11 Deg. E.W	Distance	12 Deg. N.S.	12 Deg. E.W
1	1.0	.2	1	1.0	.2	1	1.0	.2
2	2.0	.3	2	2.0	.4	2	2.0	.4
3	3.0	.5	3	2.9	.6	3	2.9	.6
4	3.9	.7	4	3.9	.8	4	3.9	.8
5	4.9	.9	5	4.9	.9	5	4.9	1.0
6	5.9	1.0	6	5.9	1.1	6	5.9	1.2
7	6.9	1.2	7	6.9	1.3	7	6.9	1.5
8	7.9	1.4	8	7.8	1.5	8	7.8	1.7
9	8.9	1.6	9	8.8	1.7	9	8.8	1.9
10	9.9	1.7	10	9.8	1.9	10	9.8	2.1
20	19.7	3.5	20	19.6	3.8	20	19.6	4.2
30	29.6	5.2	30	29.4	5.7	30	29.3	6.2
40	39.4	6.9	40	39.3	7.6	40	39.1	8.3
50	49.2	8.7	50	49.1	9.5	50	48.9	10.4
60	59.1	10.4	60	58.9	11.4	60	58.7	12.5
70	68.9	12.1	70	68.7	13.4	70	68.5	14.6
80	78.8	13.9	80	78.5	15.3	80	78.3	16.6
90	88.6	15.6	90	88.3	17.2	90	88.0	18.7
100	98.5	17.4	100	98.1	19.1	100	97.8	20.8

| Dist. | E.W. | N.S. | Dist. | E.W. | N.S. | Dist. | E.W. | N.S. |

80 Deg.   79 Deg.   78 Deg.

A 3

A Table of Northing, or Southing,

Distance	13 Deg. N.S.	E.W.	Distance	14 Deg. N.S.	E.W.	Distance	15 Deg. N.S.	E.W.
1	1.0	.2	1	1.0	.2	1	1.0	.3
2	2.0	.4	2	1.9	.5	2	1.9	.5
3	2.9	.7	3	2.9	.7	3	2.9	.8
4	3.9	.9	4	3.9	1.0	4	3.9	1.0
5	4.9	1.1	5	4.8	1.2	5	4.8	1.3
6	5.9	1.3	6	5.8	1.4	6	5.8	1.6
7	6.8	1.6	7	6.8	1.7	7	6.8	1.8
8	7.8	1.8	8	7.8	1.9	8	7.7	2.1
9	8.8	2.0	9	8.7	2.2	9	8.7	2.3
10	9.8	2.2	10	9.7	2.4	10	9.7	2.6
20	19.5	4.5	20	19.4	4.8	20	19.3	5.2
30	29.2	6.7	30	29.1	7.3	30	29.0	7.8
40	39.0	9.0	40	38.8	9.7	40	38.6	10.3
50	48.7	11.2	50	48.5	12.1	50	48.3	12.9
60	58.5	13.5	60	58.2	14.5	60	58.0	15.5
70	68.2	15.7	70	67.9	16.9	70	67.6	18.1
80	78.0	18.0	80	77.6	19.4	80	77.3	20.7
90	87.7	20.2	90	87.3	21.8	90	86.9	23.3
100	97.4	22.5	100	97.0	24.2	100	96.6	25.9

Dist.	E.W.	N.S.	Dist.	E.W.	N.S.	Dist.	E.W.	N.S.
	77 Deg.			76 Deg.			75 Deg.	

Easting, or Westing.

Distance	16 Deg. N.S.	16 Deg. E.W.	Distance	17 Deg. N.S.	17 Deg. E.W.	Distance	18 Deg. N.S.	18 Deg. E.W.
1	1.0	.3	1	1.0	.3	1	1.0	.3
2	1.9	.6	2	1.9	.6	2	1.9	.6
3	2.9	.8	3	2.9	.9	3	2.8	.9
4	3.8	1.1	4	3.8	1.2	4	3.8	1.2
5	4.8	1.4	5	4.8	1.5	5	4.7	1.5
6	5.8	1.7	6	5.7	1.7	6	5.7	1.8
7	6.7	1.9	7	6.7	2.0	7	6.6	2.2
8	7.7	2.2	8	7.6	2.3	8	7.6	2.5
9	8.6	2.5	9	8.6	2.6	9	8.5	2.8
10	9.6	2.8	10	9.6	2.9	10	9.5	3.1
20	19.2	5.5	20	19.1	5.8	20	19.0	6.2
30	28.8	8.3	30	28.7	8.8	30	28.5	9.3
40	38.4	11.0	40	38.3	11.7	40	38.0	12.4
50	48.1	13.8	50	47.8	14.6	50	47.6	15.4
60	57.7	16.5	60	57.4	17.5	60	57.1	18.5
70	67.3	19.3	70	66.9	20.5	70	66.6	21.6
80	76.9	22.0	80	76.5	23.4	80	76.1	24.7
90	86.5	24.8	90	86.1	26.3	90	85.6	27.8
100	96.1	27.6	100	95.6	29.2	100	95.1	30.9

Dist.	E.W.	N.S.	Dist.	E.W.	N.S.	Dist.	E.W.	N.S.
	74 Deg.			73 Deg.			72 Deg.	

A 4

A Table of Northing, or Southing,

Distance	19 Deg. N.S.	19 Deg. E.W.	Distance	20 Deg. N.S.	20 Deg. E.W.	Distance	21 Deg. N.S.	21 Deg. E.W.
1	.9	.3	1	.9	.3	1	.9	.4
2	1.9	.6	2	1.9	.7	2	1.9	.7
3	2.8	1.0	3	2.8	1.0	3	2.8	1.1
4	3.8	1.3	4	3.8	1.4	4	3.7	1.4
5	4.7	1.6	5	4.7	1.7	5	4.7	1.8
6	5.7	2.0	6	5.6	2.0	6	5.6	2.1
7	6.6	2.3	7	6.6	2.4	7	6.5	2.5
8	7.5	2.6	8	7.5	2.7	8	7.5	2.9
9	8.5	2.9	9	8.5	3.1	9	8.4	3.2
10	9.4	3.3	10	9.4	3.4	10	9.3	3.6
20	18.9	6.5	20	18.8	6.8	20	18.7	7.2
30	28.4	9.8	30	28.2	10.3	30	28.0	10.7
40	37.8	13.0	40	37.6	13.7	40	37.3	14.3
50	47.3	16.3	50	47.0	17.1	50	46.7	17.9
60	56.7	19.5	60	56.4	20.5	60	56.0	21.5
70	66.2	22.8	70	65.8	23.9	70	65.3	25.1
80	75.6	26.1	80	75.2	27.4	80	74.7	28.7
90	85.1	29.3	90	84.6	30.8	90	84.0	32.3
100	94.5	32.6	100	94.0	34.2	100	93.4	35.8

Dist.	E.W.	N.S.	Dist.	E.W.	N.S.	Dist.	E.W.	N.S.
	71 Deg.			70 Deg.			69 Deg.	

Easting, or Westing.

Distance	22 Deg. N.S.	22 Deg. E.W.	Distance	23 Deg. N.S.	23 Deg. E.W.	Distance	24 Deg. N.S.	24 Deg. E.W.
1	.9	.4	1	.9	.4	1	.9	.4
2	1.9	.7	2	1.8	.8	2	1.8	.8
3	2.8	1.1	3	2.8	1.2	3	2.7	1.2
4	3.7	1.5	4	3.7	1.6	4	3.6	1.6
5	4.6	1.9	5	4.6	1.9	5	4.6	2.0
6	5.6	2.2	6	5.5	2.3	6	5.5	2.4
7	6.5	2.6	7	6.4	2.7	7	6.4	2.8
8	7.4	3.0	8	7.4	3.1	8	7.3	3.2
9	8.3	3.4	9	8.3	3.5	9	8.2	3.7
10	9.3	3.7	10	9.2	3.9	10	9.1	4.1
20	18.5	7.5	20	18.4	7.8	20	18.3	8.1
30	27.8	11.2	30	27.6	11.7	30	27.4	12.2
40	37.1	15.0	40	36.8	15.6	40	36.5	16.3
50	46.4	18.7	50	46.0	19.5	50	45.7	20.3
60	55.6	22.5	60	55.2	23.4	60	54.8	24.4
70	64.9	26.2	70	64.4	27.3	70	63.9	28.5
80	74.2	30.0	80	73.6	31.2	80	73.1	32.5
90	83.4	33.7	90	82.8	35.2	90	82.2	36.6
100	92.7	37.5	100	92.0	39.1	100	91.3	40.7

| Dist. | E.W. | N.S. | Dist. | E.W. | N.S. | Dist. | E.W. | N.S. |

68 Deg.   **67 Deg.**   **66 Deg.**

A Table of Northing, or Southing,

Distance,	25 Deg. N.S.	25 Deg. E.W.	Distance,	26 Deg. N.S.	26 Deg. E.W.	Distance,	27 Deg. N.S.	27 Deg. E.W.
1	.9	.4	1	.9	.4	1	.9	.5
2	1.8	.8	2	1.8	.9	2	1.8	.9
3	2.7	1.3	3	2.7	1.3	3	2.7	1.4
4	3.6	1.7	4	3.6	1.8	4	3.6	1.8
5	4.5	2.1	5	4.5	2.2	5	4.5	2.3
6	5.4	2.5	6	5.4	2.6	6	5.3	2.7
7	6.3	3.0	7	6.3	3.1	7	6.2	3.2
8	7.2	3.4	8	7.2	3.5	8	7.1	3.6
9	8.1	3.8	9	8.1	3.9	9	8.0	4.1
10	9.1	4.2	10	9.0	4.4	10	8.9	4.5
20	18.1	8.4	20	18.0	8.8	20	17.8	9.1
30	27.2	12.7	30	27.0	13.1	30	26.7	13.6
40	36.2	16.9	40	36.0	17.5	40	35.6	18.2
50	45.3	21.1	50	44.9	21.9	50	44.5	22.7
60	54.4	25.4	60	53.9	26.3	60	53.5	27.2
70	63.4	29.6	70	62.9	30.7	70	62.4	31.8
80	72.5	33.8	80	71.9	35.1	80	71.3	36.3
90	81.6	38.0	90	80.9	39.4	90	80.2	40.9
100	90.6	42.3	100	89.9	43.8	100	89.1	45.4
Dist.	E.W.	N.S.	Dist.	E.W.	N.S.	Dist.	E.W.	N.S.
	65 Deg.			64 Deg.			63 Deg.	

Easting, or Westing.

	28 Deg.			29 Deg.			30 Deg.	
Distance,	N.S.	E.W.	Distance,	N.S.	E.W.	Distance,	N.S.	E.W.
1	.9	.5	1	.9	.5	1	.9	.5
2	1.8	.9	2	1.7	1.0	2	1.7	1.0
3	2.6	1.4	3	2.6	1.4	3	2.6	1.5
4	3.5	1.9	4	3.5	1.9	4	3.5	2.0
5	4.4	2.3	5	4.4	2.4	5	4.3	2.5
6	5.3	2.8	6	5.2	2.9	6	5.2	3.0
7	6.2	3.3	7	6.1	3.4	7	6.1	3.5
8	7.1	3.7	8	7.0	3.9	8	6.9	4.0
9	7.9	4.2	9	7.9	4.3	9	7.8	4.5
10	8.8	4.7	10	8.7	4.8	10	8.7	5.0
20	17.7	9.4	20	17.5	9.7	20	17.3	10.0
30	26.5	14.1	30	26.2	14.5	30	26.0	15.0
40	35.3	18.8	40	35.0	19.4	40	34.6	20.0
50	44.1	23.5	50	43.7	24.2	50	43.3	25.0
60	53.0	28.2	60	52.5	29.1	60	52.0	30.0
70	61.8	32.9	70	61.2	33.9	70	60.6	35.0
80	70.6	37.6	80	70.0	38.8	80	69.3	40.0
90	79.5	42.2	90	78.7	43.6	90	77.9	45.0
100	88.3	46.9	100	87.5	48.5	100	86.6	50.0

| Dist. | E.W. | N.S. | Dist. | E.W. | N.S. | Dist. | E.W. | N.S. |

| | 62 Deg. | | | 61 Deg. | | | 60 Deg. | |

A Table of Northing, or Southing,

Distance	31 Deg. N.S.	31 Deg. E.W.	Distance	32 Deg. N.S.	32 Deg. E.W.	Distance	33 Deg. N.S.	33 Deg. E.W.
1	.9	.5	1	.8	.5	1	.8	.5
2	1.7	1.0	2	1.7	1.1	2	1.7	1.1
3	2.6	1.5	3	2.5	1.6	3	2.5	1.6
4	3.4	2.1	4	3.4	2.1	4	3.4	2.2
5	4.3	2.6	5	4.2	2.6	5	4.2	2.7
6	5.1	3.1	6	5.1	3.2	6	5.0	3.3
7	6.0	3.6	7	5.9	3.7	7	5.9	3.8
8	6.9	4.1	8	6.8	4.2	8	6.7	4.4
9	7.7	4.6	9	7.6	4.8	9	7.6	4.9
10	8.6	5.1	10	8.5	5.3	10	8.4	5.4
20	17.1	10.3	20	17.0	10.6	20	16.8	10.9
30	25.7	15.4	30	25.4	15.9	30	25.2	16.3
40	34.3	20.6	40	33.9	21.2	40	33.5	21.8
50	42.9	25.7	50	42.4	26.5	50	41.9	27.2
60	51.4	30.9	60	50.9	31.8	60	50.3	32.7
70	60.0	36.0	70	59.4	37.1	70	58.7	38.1
80	68.6	41.2	80	67.8	42.4	80	67.1	43.6
90	77.1	46.3	90	76.3	47.7	90	75.5	49.0
100	85.7	51.5	100	84.8	53.0	100	83.9	54.5

Dist.	E.W.	N.S.	Dist.	E.W.	N.S.	Dist.	E.W.	N.S.
	59 Deg.			58 Deg.			57 Deg.	

Easting, or Westing.

	34 Deg.			35 Deg.			36 Deg.	
Distance	N.S.	E.W.	Distance	N.S.	E.W.	Distance	N.S.	E.W.
1	.8	.6	1	.8	.6	1	.8	.6
2	1.7	1.1	2	1.7	1.1	2	1.6	1.2
3	2.5	1.7	3	2.5	1.7	3	2.4	1.8
4	3.3	2.2	4	3.3	2.3	4	3.2	2.3
5	4.1	2.8	5	4.1	2.9	5	4.0	2.9
6	5.0	3.4	6	4.9	3.4	6	4.8	3.5
7	5.8	3.9	7	5.7	4.0	7	5.6	4.1
8	6.6	4.5	8	6.6	4.6	8	6.4	4.7
9	7.5	5.0	9	7.4	5.2	9	7.2	5.3
10	8.3	5.6	10	8.2	5.7	10	8.1	5.9
20	16.6	11.2	20	16.4	11.5	20	16.2	11.8
30	24.9	16.8	30	24.6	17.2	30	24.3	17.6
40	33.2	22.4	40	32.8	22.9	40	32.4	23.5
50	41.4	28.0	50	41.0	28.7	50	40.4	29.4
60	49.7	33.5	60	49.1	34.4	60	48.5	35.3
70	58.0	39.1	70	57.3	40.2	70	56.6	41.1
80	66.3	44.7	80	65.5	45.9	80	64.7	47.0
90	74.6	50.3	90	73.7	51.6	90	72.8	52.9
100	82.9	55.9	100	81.9	57.4	100	80.9	58.8
Dist.	E.W.	N.S.	Dist.	E.W.	N.S.	Dist.	E.W.	N.S.

| 56 Deg. | 55 Deg. | 54 Deg. |

A Table of Northing, or Southing.

Distance	37 Deg. N.S.	37 Deg. E.W.	Distance	38 Deg. N.S.	38 Deg. E.W.	Distance	39 Deg. N.S.	39 Deg. E.W.
1	.8	.6	1	.8	.6	1	.8	.6
2	1.6	1.2	2	1.6	1.2	2	1.6	1.3
3	2.4	1.8	3	2.4	1.8	3	2.3	1.9
4	3.2	2.4	4	3.1	2.5	4	3.1	2.5
5	4.0	3.0	5	3.9	3.1	5	3.9	3.1
6	4.8	3.6	6	4.7	3.7	6	4.7	3.8
7	5.6	4.2	7	5.5	4.3	7	5.4	4.4
8	6.4	4.8	8	6.3	4.9	8	6.2	5.0
9	7.2	5.4	9	7.1	5.5	9	7.0	5.7
10	8.0	6.0	10	7.9	6.2	10	7.8	6.3
20	16.0	12.0	20	15.8	12.3	20	15.5	12.6
30	24.0	18.0	30	23.6	18.5	30	23.3	18.9
40	31.9	24.1	40	31.5	24.6	40	31.1	25.2
50	39.9	30.1	50	39.4	30.8	50	38.8	31.5
60	47.9	36.1	60	47.3	36.9	60	46.6	37.8
70	55.9	42.1	70	55.2	43.1	70	54.4	44.0
80	63.9	48.1	80	63.0	49.3	80	62.2	50.3
90	71.9	54.2	90	70.9	55.4	90	69.9	56.6
100	79.9	60.2	100	78.8	61.6	100	77.7	62.9

Dist.	E.W.	N.S.	Dist.	E.W.	N.S.	Dist.	E.W.	N.S.
	53 Deg.			52 Deg.			51 Deg.	

Easting, or Westing.

Distance	40 Deg. N.S.	40 Deg. E.W.	Distance	41 Deg. N.S.	41 Deg. E.W.	Distance	42 Deg. N.S.	42 Deg. E.W.
1	.8	.6	1	.8	.7	1	.7	.7
2	1.5	1.3	2	1.5	1.3	2	1.5	1.3
3	2.3	1.9	3	2.3	2.0	3	2.2	2.0
4	3.1	2.6	4	3.0	2.6	4	3.0	2.7
5	3.8	3.2	5	3.8	3.3	5	3.7	3.3
6	4.6	3.8	6	4.5	3.9	6	4.4	4.0
7	5.4	4.5	7	5.3	4.6	7	5.2	5.7
8	6.1	5.1	8	6.0	5.2	8	5.9	5.3
9	6.9	5.8	9	6.8	5.9	9	6.7	6.0
10	7.7	6.4	10	7.5	6.6	10	7.4	6.7
20	15.3	12.9	20	15.1	13.1	20	14.9	13.4
30	23.0	19.3	30	22.6	19.7	30	22.3	20.1
40	30.6	25.7	40	30.2	26.2	40	29.7	26.8
50	38.3	32.1	50	37.7	32.8	50	37.2	33.5
60	46.0	38.6	60	45.3	39.4	60	44.6	40.1
70	53.6	45.0	70	52.8	45.9	70	52.0	46.8
80	61.3	51.4	80	60.4	52.5	80	59.4	53.5
90	68.9	57.9	90	67.9	59.0	90	66.9	60.2
100	76.6	64.3	100	75.5	65.6	100	74.3	66.9

Dist.	E.W.	N.S.	Dist.	E.W.	N.S.	Dist.	E.W.	N.S.
	50 Deg.			49 Deg.			48 Deg.	

A Table of Northing, or Southing,

Distance	43 Deg. N.S.	43 Deg. E.W.	Distance	44 Deg. N.S.	44 Deg. E.W.	Distance	45 Deg. N.S.	45 Deg. E.W.
1	.7	.7	1	.7	.7	1	.7	.7
2	1.5	1.4	2	1.4	1.4	2	1.4	1.4
3	2.2	2.0	3	2.2	2.1	3	2.1	2.1
4	2.9	2.7	4	2.9	2.8	4	2.8	2.8
5	3.6	3.4	5	3.6	3.5	5	3.5	3.5
6	4.4	4.1	6	4.3	4.2	6	4.2	4.2
7	5.1	4.8	7	5.0	4.9	7	4.9	4.9
8	5.8	5.4	8	5.8	5.6	8	5.6	5.6
9	6.6	6.1	9	6.5	6.2	9	6.4	6.4
10	7.3	6.8	10	7.2	6.9	10	7.1	7.1
20	14.6	13.6	20	14.4	13.9	20	14.1	14.1
30	21.9	20.5	30	21.6	20.8	30	21.2	21.2
40	29.2	27.3	40	28.8	27.8	40	28.3	28.3
50	36.6	34.1	50	36.0	34.7	50	35.3	35.3
60	43.9	40.9	60	43.2	41.7	60	42.4	42.4
70	51.2	47.7	70	50.3	48.6	70	49.5	49.5
80	58.5	54.6	80	57.5	55.6	80	56.6	56.6
90	65.8	61.4	90	64.7	62.5	90	63.6	63.6
100	73.1	68.2	100	71.9	69.5	100	70.7	70.7

Dist.	E.W.	N.S.	Dist.	E.W.	N.S.	Dist.	E.W.	N.S.
	47 Deg.			46 Deg.			45 Deg.	

The USE of the foregoing TABLE.

I Have already sufficiently in the 6th Chapter of this Book taught you the Use of this Table; however, because it is made somewhat different from such of this kind as have been made by others, I will briefly, by an Example or two, explain it to you. Admit in Surveying a Wood, or the like, you run a Line N E, 40 Degrees, 10 Chains: Or in plainer Terms, a Line 10 Chains in Length, that makes an Angle with the Meridian of 40 Degrees to the East-ward; and you would put down in your Field-Book the Northing and Easting of this Line, under their proper Titles, N, and E, according to Mr. *Norwood*'s way of Surveying, taught in the 6th Chapter.

First at the Head of the Table find 40 Degrees, then in the Column of Distances seek for 10 Chains: Which had, you will find to stand right against it, under the Title N, 7. 7, for the Northing, which is 7 Chains, 7/10 of a Chain: And for the Easting, under the Title E, 6. 4, which is 6 Chains, 4/10 of a Chain, as nigh as may be expressed in the Tenth Part of a Chain: But if you would know to one Link, add 0 to the Distance, so will 10 be 100, which seek for in the same Page of the Table, and right against it you will

find

find under Title N, 76, 6, or 7 Chains, 66 Links for your Northing, and under Title E, 64, 3, or 6 Chains, 43 Links for your Easting: Which found, put down in your Field-Book accordingly; and having done so by all your Lines, if you find the Northing and Southing the same, also the Easting and Westing, you may be sure you have wrought true, otherwise not.

If the Distance consists of odd Chains, and Links, as most commonly it so falls out, then take them severally out of the Table, and by adding all together you will have your Desire: As for Example;

Suppose my distance run upon any Line be N W. 35 Deg. 15 Chains, 20 Links:

First in the Table I find the Northing of 10 Chains to be

	N	
	Ch.	Lin.
10	8	19
5	4	10
20 Links 0	—	16 ¼
	12	45 ¼

which added together, makes 12 Chains 45 Links ¼ for the Northing of that distance run: In like manner under 35 Deg. and Title W, I find the Westing of the same Line, as here

	Ch.	L.
10	5	74
5	2	87
20 Links	—	11 ¼
	8	72 ¼

by which I conclude the Northing of that Line to be 12 Chains 45 Links ¼, and the Westing 8 Chains 72 Links ¼: Which thus you may prove by the Logarithms.

As

As *Radius* ——————— 10, 000000
Is to the Distance 15. 20 ——— 3, 181844
So is the Sine of the Course 35 Deg. 9, 758591

To the Westing 8 Chains 72 Links 12, 940435

And, as *Radius* ——————— 10, 000000
To the distance 15 Chains 20 Links — 3, 181844
So Co-sine of the Course 55 ——— 9, 913364

To the Northing 12 Chains 45 Links 13, 095208

Mark, if your Course had been S E, it would have been the same thing as N W: For you see in the Tables N, and S. E, and W. are joyned together. If your Degrees exceed 45, then seek for them at the Foot of the Table: And over the Titles N, S, E, W, find out the Northing, Southing, Easting or Westing.

I think this to be as much as need be said concerning the preceding Table: As for the finding the Horizontal Line of a Hill, and such like things by the Table, before you have half well read through the Chapter of *Trigonometry*, your own Ingenuity will fast enough prompt you to it.

A TABLE

OF

Sines and Tangents

To every Fifth Minute

OF THE

QUADRANT

The Table of Sines and Tangents.

0.

M.	Sine,	Co-sine.	Tangent.	Co-tangent.	
0	0.000000	10.000000	0.000000	Infinita.	60
5	7.162696	9.999999	7.162696	12.837304	55
10	7.463726	9.999998	7.463727	12.536273	50
15	7.639816	9.999996	7.639820	12.360180	45
20	7.764754	9.999993	7.764761	12.235239	40
25	7.861662	9.999989	7.861674	12.138326	35
30	7.940847	9.999988	7.940858	12.059142	30
35	8.007787	9.999977	8.007809	11.992191	25
40	8.067776	9.999971	8.065806	11.934194	20
45	8.116926	9.999963	8.116963	11.883037	15
50	8.162681	9.999954	8.162737	11.837273	10
55	8.204070	9.999944	8.204126	11.795874	5
60	8.241855	9.999934	8.241921	11.758079	0
	Co-sine.	Sine.	Co-tang.	Tangent.	M.

89.

1.

M.	Sine.	Co-sine.	Tangent.	Co-tangent.	
0	8.241855	9.999934	8.241921	11.758079	60
5	8.276614	9.999922	8.276691	11.723309	55
10	8.308794	9.999910	8.308884	11.691116	50
15	8.338753	9.999897	8.338856	11.661144	45
20	8.366777	9.999882	8.366895	11.633105	40
25	8.393101	9.999867	8.393234	11.606766	35
30	8.417919	9.999851	8.418068	11.581932	30
35	8.441394	9.999834	8.441560	11.458440	25
40	8.463665	9.999816	8.463849	11.536151	20
45	8.484848	9.999797	8.485050	11.514950	15
50	8.505045	9.999778	8.505267	11.494733	10
55	8.524343	9.999757	8.524586	11.475414	5
60	8.542810	9.999735	8.543084	11.456916	0
	Co-sine.	Sine.	Co-tang.	Tangent.	M.

88.

The Table of Sines and Tangents.

2.

M.	Sine.	Co-sine.	Tangent.	Co-tangent.	
0	8.542819	9.999735	8.543084	11.456916	60
5	8.560540	9.999713	8.560828	11.439172	55
10	8.577566	9.999689	8.577877	11.422123	50
15	8.593948	9.999665	8.594283	11.405717	45
20	8.609734	9.999640	8.610094	11.389906	40
25	8.624965	9.999614	8.625352	11.374648	35
30	8.639680	9.999586	8.640093	11.359907	30
35	8.653911	9.999558	8.654352	11.345648	25
40	8.667689	9.999529	8.668160	11.331840	20
45	8.681043	9.999500	8.681544	11.318456	15
50	8.693998	9.999469	8.694529	11.305471	10
55	8.706577	9.999437	8.707140	11.292860	5
60	8.718800	9.999404	8.719396	11.280604	0
	Co-fine.	Sine.	Co-tang.	Tangent.	M.

87.

3.

M.	Sine.	Co-sine.	Tangent.	Co-tangent.	
0	8.718800	9.999404	8.719396	11.280604	60
5	8.730688	9.999371	8.731317	11.268683	55
10	8.742250	9.999336	8.742922	11.257078	50
15	8.753528	9.999301	8.754227	11.245773	45
20	8.764511	9.999265	8.765246	11.234754	40
25	8.775223	9.999227	8.775995	11.224005	35
30	8.785675	9.999189	8.786486	11.213514	30
35	8.795881	9.999150	8.796731	11.203269	25
40	8.805852	9.999110	8.806742	11.193258	20
45	8.815599	9.999069	8.816529	11.183471	15
50	8.825130	9.999027	8.826103	11.173897	10
55	8.834456	9.998984	8.835471	11.164529	5
60	8.843585	9.998941	8.844644	11.155356	0
	Co-fine.	Sine.	Co-tang.	Tangent.	M.

86.   B 4

The Table of Sines and Tangents.

4.

M.	Sine.	Co-fine.	Tangent.	Co-tangent.	
0	8.843585	9.998941	8.844644	11.155356	60
5	8.852525	9.998896	8.853628	11.146372	55
10	8.861283	9.998851	8.862433	11.137567	50
15	8.869868	9.998804	8.871064	11.128936	45
20	8.878285	9.998757	8.879529	11.120471	40
25	8.886542	9.998708	8.887833	11.112167	35
30	8.894643	9.998659	8.895984	11.104016	30
35	8.902596	9.998609	8.903987	11.096013	25
40	8.910414	9.998558	8.911846	11.088154	20
45	8.918073	9.998506	8.919568	11.080432	15
50	8.925609	9.998453	8.927156	11.072844	10
55	8.933015	9.998399	8.934616	11.065384	5
60	8.940296	9.998344	8.941952	11.058048	0
	Co-fine.	Sine.	Co-tang.	Tangent.	M.

85.

5.

M.	Sine.	Co-fine.	Tangent.	Co-tangent.	
0	8.940296	9.998344	8.941952	11.058048	60
5	8.947456	9.998289	8.949168	11.050832	55
10	8.954499	9.998232	8.956267	11.043733	50
15	8.961429	9.998174	8.963255	11.036745	45
20	8.968249	9.998116	8.970133	11.029867	40
25	8.974962	9.998056	8.976906	11.023094	35
30	8.981573	9.997996	8.983577	11.016423	30
35	8.988083	9.997935	8.990149	11.009851	25
40	8.994497	9.997872	8.996624	11.003376	20
45	9.000816	9.997809	9.003007	10.996993	15
50	9.007044	9.997745	9.009298	10.990702	10
55	9.013182	9.997680	9.015502	10.984498	5
60	9.019235	9.997614	9.021620	10.978380	0
	Co-fine.	Sine.	Co-tang.	Tangent.	M.

84.

The Table of Sines and Tangents.

6.

M.	Sine.	Co-sine.	Tangent.	Co-tangent.	
0	9.019235	9.997614	9.021620	10.978380	60
5	9.025203	9.997547	9.027655	10.972345	55
10	9.031089	9.997480	9.033609	10.966391	50
15	9.036896	9.997411	9.039485	10.960515	45
20	9.042625	9.997341	9.045284	10.954716	40
25	9.048279	9.997271	9.051008	10.948992	35
30	9.053859	9.997199	9.056659	10.943341	30
35	9.059367	9.997127	9.062240	10.937760	25
40	9.064806	9.997053	9.067752	10.932248	20
45	9.070176	9.996979	9.073197	10.926803	15
50	9.075480	9.996904	9.078576	10.921424	10
55	9.080719	9.996828	9.083891	10.916109	5
60	9.085894	9.996751	9.089144	10.910856	0
	Co-sine.	Sine.	Co-tang.	Tangent.	M.

83.

7.

M.	Sine.	Co-sine.	Tangent.	Co-tangent.	
0	9.085894	9.996751	9.089144	10.910850	60
5	9.091008	9.996673	9.094336	10.905664	55
10	9.096062	9.996594	9.099468	10.900532	50
15	9.101056	9.996514	9.104542	10.895458	45
20	9.105992	9.996433	9.109559	10.890441	40
25	9.110873	9.996351	9.114521	10.885479	35
30	9.115698	9.996269	9.119429	10.880571	30
35	9.120469	9.996185	9.124284	10.875716	25
40	9.125187	9.996100	9.129087	10.870913	20
45	9.129854	9.996015	9.133839	10.866161	15
50	9.134470	9.995928	9.138542	10.861458	10
55	9.139037	9.995841	9.143196	10.856804	5
60	9.143555	9.995753	9.147803	10.852197	0
	Co-sine.	Sine.	Co-tang.	Tangent.	M.

82.

The Table of Sines and Tangents.

8.

M.	Sine.	Co-sine.	Tangent.	Co-tangent.	
0	9.143555	9.995753	9.147803	10.852197	60
5	9.148026	9.995664	9.152363	10.847637	55
10	9.152451	9.995573	9.156877	10.843123	50
15	9.156830	9.995482	9.161347	10.838653	45
20	9.161164	9.995390	9.165774	10.834226	40
25	9.165454	9.995297	9.170157	10.829843	35
30	9.169702	9.995203	9.174499	10.825501	30
35	9.173908	9.995108	9.178799	10.821201	25
40	9.178072	9.995013	9.183059	10.816941	20
45	9.182196	9.994916	9.187280	10.812720	15
50	9.186280	9.994818	9.191462	10.808538	10
55	9.190325	9.994720	9.195606	10.804394	5
60	9.194332	9.994620	9.199713	10.800287	0
	Co-sine.	Sine.	Co-tang.	Tangent.	M.

81.

9.

M.	Sine.	Co-sine.	Tangent.	Co-tangent.	
0	9.194332	9.994620	9.199713	10.800287	60
5	9.198302	9.994519	9.203782	10.796218	55
10	9.202234	9.994418	9.207817	10.792183	50
15	9.206131	9.994316	9.211815	10.788185	45
20	9.209992	9.994212	9.215780	10.784220	40
25	9.213818	9.994108	9.219710	10.780290	35
30	9.217609	9.994003	9.223607	10.776393	30
35	9.221367	9.993897	9.227471	10.772529	25
40	9.225092	9.993789	9.231302	10.768698	20
45	9.228784	9.993681	9.235103	10.764897	15
50	9.232444	9.993572	9.238872	10.761128	10
55	9.236073	9.993462	9.242610	10.757390	5
60	9.239670	9.993351	9.246319	10.753681	0
	Co-sine.	Sine.	Co-tang.	Tangent.	M.

80

The Table of Sines and Tangents.

10.

M.	Sine.	Co-sine.	Tangent.	Co-tangent.	
0	9.239670	9.993351	9.246319	10.753681	60
5	9.243237	9.993240	9.249998	10.750002	55
10	9.246775	9.993127	9.253648	10.746352	50
15	9.250282	9.993013	9.257269	10.742731	45
20	9.253761	9.992898	9.260863	10.739137	40
25	9.257211	9.992783	9.264428	10.735572	35
30	9.260633	9.991666	9.267967	10.732033	30
35	9.264027	9.992549	9.271479	10.728521	25
40	9.267395	9.992430	9.274964	10.725036	20
45	9.270735	9.992311	9.278424	10.721576	15
50	9.274049	9.992190	9.281858	10.718142	10
55	9.277337	9.992069	9.285268	10.714732	5
60	9.280599	9.991947	9.288652	10.711348	0
	Co-sine.	Sine.	Co-tang.	Tangent.	M.

79.

11.

M.	Sine.	Co-sine.	Tangent.	Co-tangent.	
0	9.280599	9.991974	9.288652	10.711348	60
5	9.283836	9.991823	9.292013	10.707987	55
10	9.287048	9.991699	9.295349	10.704651	50
15	9.290236	9.991574	9.298662	10.701338	45
20	9.293399	9.991448	9.301951	10.698049	40
25	9.296539	9.991321	9.305218	10.694782	35
30	9.299655	9.991193	9.308463	10.691537	30
35	9.302748	9.991064	9.311685	10.688315	25
40	9.305819	9.990934	9.314885	10.685115	20
45	9.308867	9.990803	9.318064	10.681936	15
50	9.311893	9.990671	9.321222	10.678778	10
55	9.314897	9.990538	9.324358	10.675642	5
60	9.317879	9.990404	9.327475	10.672525	0
	Co-sine.	Sine.	Co-tang.	Tangent	M.

78.

The Table of Sines and Tangents.

12.

M.	Sine.	Co-sine.	Tangent.	Co-tangent.	
0	9.317879	9.990404	9.327475	10.672525	60
5	9.320840	9.990270	9.330570	10.669430	55
10	9.323780	9.990134	9.333646	10.666354	50
15	9.326700	9.989997	9.336702	10.663298	45
20	9.329599	9.989860	9.339739	10.660261	40
25	9.332478	9.989721	9.342757	10.657243	35
30	9.335337	9.989582	9.345755	10.654245	30
35	9.338176	9.989441	9.348735	10.651265	25
40	9.340996	9.989300	9.351697	10.648303	20
45	9.343797	9.989157	9.354640	10.645360	15
50	9.346579	9.989014	9.357566	10.642434	10
55	9.349343	9.988869	9.360474	10.639526	5
60	9.352088	9.988724	9.363364	10.636636	0
	Co-sine.	Sine.	Co-tang.	Tangent.	M.

77.

13.

M.	Sine.	Co-sine.	Tangent.	Co-tangent.	
0	9.352088	9.988724	9.363364	10.636636	60
5	9.354815	9.988578	9.366237	10.633763	55
10	9.357524	9.988430	9.369094	10.630906	50
15	9.360215	9.988282	9.371933	10.628067	45
20	9.362889	9.988133	9.374756	10.625244	40
25	9.365546	9.987983	9.377563	10.622437	35
30	9.368185	9.987832	9.380354	10.619646	30
35	9.370808	9.987679	9.383129	10.616871	25
40	9.373414	9.987526	9.385888	10.614112	20
45	9.376003	9.987372	9.388631	10.611369	15
50	9.378577	9.987217	9.391360	10.608640	10
55	9.381134	9.987061	9.394073	10.605927	5
60	9.383675	9.986904	9.396771	10.603229	0
	Co-sine.	Sine.	Co-tang.	Tangent.	M.

76.

The Table of Sines and Tangents.

14.

M.	Sine.	Co-sine.	Tangent.	Co-tangent.	
0	9.383675	9.986904	9.396771	10.603229	60
5	9.386201	9.986746	9.399455	10.600545	55
10	9.388711	9.986587	9.402124	10.597876	50
15	9.391206	9.986427	9.404778	10.595222	45
20	9.393685	9.986266	9.407419	10.592581	40
25	9.396150	9.986104	9.410045	10.589955	35
30	9.398600	9.985942	9.412658	10.587342	30
35	9.401035	9.985778	9.415257	10.584743	25
40	9.403455	9.985613	9.417842	10.582158	20
45	9.405862	9.985447	9.420415	10.579585	15
50	9.408254	9.985280	9.422974	10.577026	10
55	9.410632	9.985113	9.425519	10.574481	5
60	9.412996	9.984944	9.428052	10.571948	0
	Co-sine.	Sine.	Co-tang.	Tangent.	M.

75.

15.

M.	Sine.	Co-sine.	Tangent.	Co-tangent.	
0	9.412996	9.984944	9.428052	10.571948	60
5	9.415347	9.984774	9.430573	10.569427	55
10	9.417684	9.984603	9.433080	10.566920	50
15	9.420007	9.984432	9.435576	10.564424	45
20	9.422318	9.984259	9.438059	10.561941	40
25	9.424615	9.984085	9.440529	10.559471	35
30	9.426899	9.983911	9.442988	10.557012	30
35	9.429170	9.983735	9.445435	10.554565	25
40	9.431429	9.983558	9.447870	10.552130	20
45	9.433675	9.983381	9.450294	10.549706	15
50	9.435908	9.983202	9.452706	10.547294	10
55	9.438129	9.983022	9.455107	10.544893	5
60	9.440338	9.982842	9.457496	10.542504	0
	Co-sine.	Sine.	Co-tang.	Tangent.	M.

74.

The Table of Sines and Tangents.

16.

M.	Sine.	Co-sine.	Tangent.	Co-tangent.	
0	9.440338	9.982842	9.457496	10.542504	60
5	9.442535	9.982660	9.459875	10.540125	55
10	9.444720	9.982477	9.462242	10.537758	50
15	9.446893	9.982294	9.464599	10.535401	45
20	9.449054	9.982109	9.466945	10.533055	40
25	9.451204	9.981924	9.469280	10.530720	35
30	9.453342	9.981737	9.471605	10.528395	30
35	9.455469	9.981549	9.473919	10.526081	25
40	9.457584	9.981361	9.476223	10.523777	20
45	9.459688	9.981171	9.478517	10.521483	15
50	9.461782	9.980981	9.480801	10.519199	10
55	9.463864	9.980789	9.483075	10.516925	5
60	9.465935	9.980596	9.485339	10.514661	0
	Co-sine.	Sine.	Co-tang.	Tangent.	M

73.

17.

M.	Sine.	Co-sine.	Tangent.	Co-tangent.	
0	9.465935	9.980596	9.485339	10.514661	60
5	9.467996	9.980403	9.487593	10.512407	55
10	9.470446	9.980208	9.489838	10.510162	50
15	9.472086	9.980012	9.492073	10.507927	45
20	9.474115	9.979816	9.494299	10.505701	40
25	9.476133	9.979618	9.496515	10.503485	35
30	9.478142	9.979420	9.498722	10.501278	30
35	9.480140	9.979220	9.500920	10.499080	25
40	9.482128	9.979019	9.503109	10.496891	20
45	9.484107	9.978817	9.505289	10.494711	15
50	9.486075	9.978615	9.507460	10.492540	10
55	9.488034	9.978411	9.509622	10.490378	5
60	9.489982	9.978206	9.511776	10.488224	0
	Co-sine.	Sine.	Co-tang.	Tangent.	M

72.

The Table of Sines and Tangents.

18.

M.	Sine.	Co-sine.	Tangent.	Co-tangent.	
0	9.489982	9.978206	9.511776	10.488224	60
5	9.491922	9.978001	9.513921	10.486079	55
10	9.493851	9.977794	9.516057	10.483943	50
15	9.495772	9.977586	9.518186	10.481814	45
20	9.497682	9.977377	9.520305	10.479695	40
25	9.499584	9.977167	9.522417	10.477583	35
30	9.501476	9.976957	9.524520	10.475480	30
35	9.503360	9.976745	9.526615	10.473385	25
40	9.505234	9.976532	9.528702	10.471298	20
45	9.507099	9.976318	9.530781	10.469219	15
50	9.508956	9.976103	9.532853	10.467147	10
55	9.510803	9.975887	9.534916	10.465084	5
60	9.512642	9.975670	9.536972	10.463028	0
	Co-sine.	Sine.	Co-tang.	Tangent.	M.

71.

19.

M.	Sine.	Co-sine.	Tangent.	Co-tangent.	
0	9.512642	9.975670	9.536972	10.463028	60
5	9.514472	9.975452	9.539020	10.460980	55
10	9.516294	9.975233	9.541061	10.458939	50
15	9.518107	9.975013	9.543094	10.456906	45
20	9.519911	9.974792	9.545119	10.454881	40
25	9.521707	9.974570	9.547138	10.452862	35
30	9.523495	9.974347	9.549149	10.450851	30
35	9.525275	9.974122	9.551153	10.448847	25
40	9.527046	9.973897	9.553149	10.446851	20
45	9.528810	9.973671	9.555139	10.444861	15
50	9.530565	9.973444	9.557121	10.442879	10
55	9.532312	9.973215	9.559097	10.440903	5
60	9.534052	9.972986	9.561066	10.438934	0
	Co-sine.	Sine.	Co-tang.	Tangent.	M.

70.

The Table of Sines and Tangents.

20.

M.	Sine.	Co-sine.	Tangent.	Co-tangent.	
0	9.534052	9.972986	9.561066	10.438934	60
5	9.535783	9.972755	9.563028	10.436972	55
10	9.537507	9.972524	9.564983	10.435017	50
15	9.539223	9.972291	9.566932	10.433068	45
20	9.540931	9.972058	9.568873	10.431127	40
25	9.542632	9.971823	9.570809	10.429191	35
30	9.544325	9.971583	9.572738	10.427262	30
35	9.546011	9.971351	9.574660	10.425340	25
40	9.547689	9.971113	9.576576	10.423424	20
45	9.549360	9.970874	9.578486	10.421514	15
50	9.551024	9.970635	9.580389	10.419611	10
55	9.552680	9.970394	9.582286	10.417714	5
60	9.554329	9.970152	9.584177	10.415823	0
	Co-sine.	Sine.	Co-tang.	Tangent.	M.

69.

21.

M.	Sine.	Co-sine.	Tangent.	Co-tangent.	
0	9.554329	9.970152	9.584177	10.415823	60
5	9.555971	9.969909	9.586062	10.413938	55
10	9.557606	9.969665	9.587941	10.412059	50
15	9.559234	9.969420	9.589814	10.410186	45
20	9.560855	9.969173	9.591681	10.408319	40
25	9.562468	9.968926	9.593542	10.406458	35
30	9.564075	9.968678	9.595398	10.404602	30
35	9.565676	9.968429	9.597247	10.402753	25
40	9.567269	9.968178	9.599091	10.400909	20
45	9.568856	9.967927	9.600929	10.399071	15
50	9.570435	9.967674	9.602761	10.397239	10
55	9.572009	9.967421	9.604588	10.395412	5
60	9.573575	9.967166	9.606410	10.393590	0
	Co-sine.	Sine.	Co-tang.	Tangent.	M.

68.

The Table of Sines and Tangents.

22.

M.	Sine.	Co-sine.	Tangent.	Co-tangent.	
0	9.573575	9.967166	9.606410	10.393590	60
5	9.575136	9.966910	9.608225	10.391775	55
10	9.576689	9.966653	9.610036	10.389964	50
15	9.578236	9.966395	9.611841	10.388159	45
20	9.579777	9.966136	9.613641	10.386359	40
25	9.581312	9.965876	9.615435	10.384565	35
30	9.582840	9.965615	9.617224	10.382776	30
35	9.584361	9.965353	9.619008	10.380992	25
40	9.585877	9.965090	9.620787	10.379213	20
45	9.587386	9.964826	9.622561	10.377439	15
50	9.588890	9.964560	9.624330	10.375670	10
55	9.590387	9.964294	9.626093	10.373907	5
60	9.591878	9.964026	9.627852	10.372148	0
	Co-fine.	Sine.	Co-tang.	Tangent.	M.

67.

23.

M.	Sine.	Co-sine.	Tangent.	Co-tangent.	
0	9.591878	9.964026	9.627852	10.372148	60
5	9.593363	9.963757	9.629606	10.370394	55
10	9.594842	9.963488	9.631355	10.368645	50
15	9.596315	9.963217	9.633098	10.366902	45
20	9.597783	9.962945	9.634838	10.365162	40
25	9.599244	9.962672	9.636572	10.363428	35
30	9.600700	9.962398	9.638302	10.361698	30
35	9.602150	9.962123	9.640027	10.359973	25
40	9.603594	9.961846	9.641747	10.358253	20
45	9.605032	9.961569	9.643463	10.356537	15
50	9.606465	9.961290	9.645174	10.354826	10
55	9.607892	9.961011	9.646881	10.353119	5
60	9.609313	9.960730	9.648583	10.351417	0
	Co-fine.	Sine.	Co-tang.	Tangent.	M.

66.

The Table of Sines and Tangents.

24.

M.	Sine.	Co-sine.	Tangent.	Co-tangent.	
0	9.609313	9.960730	9.648583	10.351417	60
5	9.610729	9.960448	9.650281	10.349719	55
10	9.612140	9.960165	9.651974	10.348026	50
15	9.613545	9.959882	9.653663	10.346337	45
20	9.614944	9.959596	9.655348	10.344652	40
25	9.616338	9.959310	9.657028	10.342972	35
30	9.617727	9.959023	9.658704	10.341296	30
35	9.619110	9.958734	9.660376	10.339624	25
40	9.620488	9.958445	9.662043	10.337957	20
45	9.621861	9.958154	9.663707	10.336293	15
50	9.623229	9.957863	9.665366	10.334634	10
55	9.624591	9.957570	9.667021	10.332979	5
60	9.625948	9.957276	9.668673	10.331327	0
	Co-sine.	Sine.	Co-tang.	Tangent.	M.

65.

25.

M.	Sine.	Co-sine.	Tangent.	Co-tangent.	
0	9.625948	9.957276	9.668673	10.331327	60
5	9.627300	9.956981	9.670320	10.329680	55
10	9.628647	9.956684	9.671963	10.328037	50
15	9.629989	9.956387	9.673602	10.326398	45
20	9.631326	9.956089	9.675237	10.324763	40
25	9.632658	9.955789	9.676869	10.323131	35
30	9.633984	9.955488	9.678496	10.322504	30
35	9.635306	9.955186	9.680120	10.319880	25
40	9.636623	9.954883	9.681740	10.318260	20
45	9.637935	9.954579	9.683356	10.316644	15
50	9.639242	9.954274	9.684968	10.315032	10
55	9.640544	9.953968	9.686577	10.313423	5
60	9.641842	9.953660	9.688182	10.311818	0
	Co-sine.	Sine.	Co-tang.	Tangent.	M.

64

The Table of Sines and Tangents.

26.

M.	Sine.	Co-sine.	Tangent.	Co-tangent.	
0	9.641842	9.953660	9.688182	10.311818	60
5	9.643135	9.953352	9.689783	10.310217	55
10	9.644423	9.953042	9.691381	10.308619	50
15	9.645706	9.952731	9.692975	10.307025	45
20	9.646984	9.952419	9.694566	10.305434	40
25	9.648258	9.952106	9.696153	10.303847	35
30	9.649527	9.951791	9.697736	10.302264	30
35	9.650792	9.951476	9.699316	10.300684	25
40	9.652052	9.951159	9.700893	10.299107	20
45	9.653308	9.950841	9.702466	10.297534	15
50	9.654558	9.950522	9.704036	10.295964	10
55	9.655805	9.950202	9.705603	10.294397	5
60	9.657047	9.949881	9.707166	10.292834	0
	Co-sine.	Sine.	Co-tang.	Tangent.	M.

63.

27.

M.	Sine.	Co-sine.	Tangent.	Co-tangent.	
0	9.657047	9.949881	9.707166	10.292834	60
5	9.658284	9.949558	9.708726	10.291274	55
10	9.659517	9.949235	9.710282	10.289718	50
15	9.660746	9.948910	9.711836	10.288104	45
20	9.661970	9.948584	9.713386	10.286614	40
25	9.663190	9.948257	9.714933	10.285067	35
30	9.664406	9.947929	9.716477	10.283523	30
35	9.665617	9.947600	9.718017	10.281983	25
40	9.666824	9.947269	9.719555	10.280445	20
45	9.668027	9.946937	9.721089	10.278911	15
50	9.669225	9.946604	9.722621	10.277379	10
55	9.670419	9.946270	9.724149	10.275851	5
60	9.671609	9.945935	9.725674	10.274326	0
	Co-sine.	Sine.	Co-tang.	Tangent.	M.

62.

The Table of Sines and Tangents.

28.

M.	Sine.	Co-sine.	Tangent.	Co-tangent.	
0	9.671609	9.945935	9.725674	10.274326	60
5	9.672795	9.945598	9.727197	10.272803	55
10	9.673977	9.945261	9.728716	10.271284	50
15	9.675155	9.944922	9.730233	10.269767	45
20	9.676328	9.944582	9.731746	10.268254	40
25	9.677498	9.944241	9.733257	10.266743	35
30	9.678663	9.943899	9.734764	10.265236	30
35	9.679824	9.943555	9.736269	10.263731	25
40	9.680982	9.943210	9.737771	10.262229	20
45	9.682135	9.942864	9.739271	10.260729	15
50	9.683284	9.942517	9.740767	10.259233	10
55	9.684430	9.942169	9.742261	10.257739	5
60	9.685571	9.941819	9.743752	10.256248	0
	Co-sine.	Sine.	Co-tang.	Tangent.	M.

61.

29.

M.	Sine.	Co-sine.	Tangent.	Co-tangent.	
0	9.685571	9.941819	9.743751	10.256248	60
5	9.686709	9.941469	9.745240	10.254760	55
10	9.687843	9.941117	9.746726	10.253274	50
15	9.688972	9.940763	9.748209	10.251791	45
20	9.690098	9.940409	9.749689	10.250311	40
25	9.691220	9.940054	9.751167	10.248833	35
30	9.692339	9.939697	9.752642	10.247358	30
35	9.693453	9.939339	9.754115	10.245885	25
40	9.694564	9.938980	9.755585	10.244415	20
45	9.695671	9.938619	9.757052	10.242948	15
50	9.696775	9.938258	9.758517	10.241483	10
55	9.697874	9.937895	9.759979	10.240021	5
60	9.698970	9.937531	9.761439	10.238561	0
	Co-sine.	Sine.	Co-tang.	Tangent.	M.

60.

The Table of Sines and Tangents.

30.

M.	Sine.	Co-sine.	Tangent.	Co-tangent.	
0	9.698970	9.937531	9.761439	10.238561	60
5	9.700062	9.937165	9.762897	10.237103	55
10	9.701151	9.936799	9.764352	10.235648	50
15	9.702236	9.936431	9.765805	10.234195	45
20	9.703317	9.936062	9.767255	10.232745	40
25	9.704395	9.935692	9.768703	10.231297	35
30	9.705469	9.935320	9.770148	10.229852	30
35	9.706539	9.934948	9.771592	10.228408	25
40	9.707606	9.934574	9.773033	10.226967	20
45	9.708670	9.934199	9.774471	10.225529	15
50	9.709730	9.933822	9.775908	10.224092	10
55	9.710786	9.933445	9.777342	10.222658	5
60	9.711839	9.933066	9.778774	10.221226	0
	Co-fine.	Sine.	Co-tang.	Tangent.	M.

59.

31.

M.	Sine.	Co-sine.	Tangent.	Co-tangent.	
0	9.711839	9.933066	9.778774	10.221226	60
5	9.712889	9.932685	9.780203	10.219797	55
10	9.713935	9.932304	9.781631	10.218369	50
15	9.714978	9.931921	9.783056	10.216944	45
20	9.716017	9.931537	9.784479	10.215521	40
25	9.717053	9.931152	9.785900	10.214100	35
30	9.718085	9.930766	9.787319	10.212681	30
35	9.719114	9.930378	9.788736	10.211264	25
40	9.720140	9.929989	9.790151	10.209849	20
45	9.721162	9.929599	9.791563	10.208437	15
50	9.722181	9.929207	9.792974	10.207026	10
55	9.723197	9.928815	9.794383	10.205617	5
60	9.724210	9.928420	9.795789	10.204211	0
	Co-fine.	Sine.	Co-tang.	Tangent.	M.

The Table of Sines and Tangents.

32.

M.	Sine.	Co-sine.	Tangent.	Co-tangent.	
0	9.724210	9.928420	9.795789	10.204211	60
5	9.725219	9.928025	9.797194	10.202806	55
10	9.726225	9.927629	9.798596	10.201404	50
15	9.727228	9.927231	9.799997	10.200003	45
20	9.728227	9.926831	9.801396	10.198604	40
25	9.729223	9.926431	9.802792	10.197208	35
30	9.730217	9.926029	9.804187	10.195813	30
35	9.731206	9.925626	9.805580	10.194420	25
40	9.732193	9.925222	9.806971	10.193029	20
45	9.733177	9.924816	9.808361	10.191639	15
50	9.734157	9.924409	9.809748	10.190252	10
55	9.735135	9.924001	9.811134	10.188866	5
60	9.736109	9.923591	9.812517	10.187483	0
	Co-sine.	Sine.	Co-tang.	Tangent.	M.

57.

33.

M.	Sine.	Co-sine.	Tangent.	Co-tangent.	
0	9.736109	9.923591	9.812517	10.187483	60
5	9.737080	9.923181	9.813899	10.186101	55
10	9.738048	9.922769	9.815280	10.184720	50
15	9.739013	9.922355	9.816658	10.183342	45
20	9.739975	9.921940	9.818035	10.181965	40
25	9.740934	9.921524	9.819410	10.180590	35
30	9.741889	9.921107	9.820783	10.179217	30
35	9.742842	9.920688	9.822154	10.177846	25
40	9.743792	9.920268	9.823524	10.176476	20
45	9.744739	9.919846	9.824893	10.175107	15
50	9.745683	9.919424	9.826259	10.173741	10
55	9.746624	9.919000	9.827624	10.172376	5
60	9.747562	9.918574	9.828987	10.171013	0
	Co-sine.	Sine.	Co-tang.	Tangent.	M.

56.

The Table of Sines and Tangents.

34.

M.	Sine.	Co-sine.	Tangent.	Co-tangent.	
0	9.747562	9.918574	9.828987	10.171013	60
5	9.748497	9.918147	9.830349	10.169651	55
10	9.749429	9.917719	9.831709	10.168291	50
15	9.750358	9.917290	9.833068	10.166932	45
20	9.751284	9.916859	9.834425	10.165575	40
25	9.752208	9.916427	9.835780	10.164220	35
30	9.753128	9.915994	9.837134	10.162866	30
35	9.754046	9.915559	9.838487	10.161513	25
40	9.754960	9.915123	9.839838	10.160162	20
45	9.755872	9.914685	9.841187	10.158813	15
50	9.756782	9.914246	9.842535	10.157465	10
55	9.757688	9.913806	9.843882	10.156118	5
60	9.758591	9.913365	9.845227	10.154773	0
	Co-sine.	Sine.	Co.tang.	Tangent.	M.

55.

35.

M.	Sine.	Co-sine.	Tangent.	Co-tangent.	
0	9.758591	9.913365	9.845227	10.154773	60
5	9.759492	9.912922	9.846570	10.153430	55
10	9.760390	9.912477	9.847913	10.152087	50
15	9.761285	9.912031	9.849254	10.150746	45
20	9.762177	9.911584	9.850593	10.149407	40
25	9.763067	9.911136	9.851931	10.148069	35
30	9.763954	9.910686	9.853268	10.146732	30
35	9.764838	9.910235	9.854603	10.145397	25
40	9.765720	9.909782	9.855938	10.144062	20
45	9.766598	9.909328	9.857270	10.142730	15
50	9.767475	9.908873	9.858602	10.141398	10
55	9.768348	9.908416	9.859932	10.140068	5
60	9.769219	9.907958	9.861261	10.138739	0
	Co-sine.	Sine.	Co-tang.	Tangent.	M.

54.

The Table of Sines and Tangents.

36.

M.	Sine.	Co-fine.	Tangent.	Co-tangent.	
0	9.769219	9.907758	9.861261	10.138739	60
5	9.770087	9.907498	9.862589	10.137411	55
10	9.770952	9.907037	9.863915	10.136085	50
15	9.771815	9.906575	9.865240	10.134760	45
20	9.772675	9.906111	9.866564	10.133436	40
25	9.773533	9.905645	9.867887	10.132133	35
30	9.774388	9.905179	9.869209	10.130791	30
35	9.775240	9.904711	9.870529	10.129471	25
40	9.776090	9.904241	9.871849	10.128151	20
45	9.776937	9.903770	9.873167	10.126833	15
50	9.777781	9.903298	9.874484	10.125516	10
55	9.778624	9.902824	9.875800	10.124200	5
60	9.779463	9.902349	9.877114	10.122886	0
	Co-fine.	Sine.	Co-tang.	Tangent.	M.

53.

37.

M.	Sine.	Co-fine.	Tangent.	Co-tangent.	
0	9.779463	9.902349	9.877114	10.122886	60
5	9.780300	9.901872	9.878428	10.121572	55
10	9.781134	9.901394	9.879741	10.120259	50
15	9.781966	9.900914	9.881052	10.118948	45
20	9.782796	9.900433	9.882363	10.117637	40
25	9.783623	9.899951	9.883672	10.116328	35
30	9.784447	9.899467	9.884980	10.115020	30
35	9.785269	9.898981	9.886288	10.113712	25
40	9.786089	9.898494	9.887594	10.112406	20
45	9.786906	9.898006	9.888900	10.111100	15
50	9.787720	9.897516	9.890204	10.109796	10
55	9.788532	9.897025	9.891507	10.108493	5
60	9.789342	9.896530	9.892810	10.107190	0
	Co-fine.	Sine.	Co-tang.	Tangent.	M.

52.

The Table of Sines and Tangents.

38.

M.	Sine.	Co-sine.	Tangent.	Co-tangent.	
0	9.789342	9.896532	9.892810	10.107190	60
5	9.790149	9.896038	9.894111	10.105889	55
10	9.790954	9.895542	9.895412	10.104588	50
15	9.791757	9.895045	9.896712	10.103288	45
20	9.792557	9.894546	9.898010	10.101990	40
25	9.793354	9.894046	9.899308	10.100692	35
30	9.794150	9.893544	9.900605	10.099395	30
35	9.794942	9.893041	9.901901	10.098099	25
40	9.795733	9.892536	9.903197	10.096803	20
45	9.796521	9.892030	9.904491	10.095509	15
50	9.797307	9.891523	9.905785	10.094215	10
55	9.798091	9.891013	9.907077	10.092923	5
60	9.798872	9.890503	9.908369	10.091631	0
	Co-fine.	Sine.	Co-tang.	Tangent.	M.

51.

39.

M.	Sine.	Co-sine.	Tangent.	Co-tangent.	
0	9.798872	9.890503	9.908369	10.091631	60
5	9.799651	9.889990	9.909660	10.090340	55
10	9.800427	9.889477	9.910951	10.089049	50
15	9.801201	9.888961	9.912240	10.087760	45
20	9.801973	9.888444	9.913529	10.086471	40
25	9.802743	9.887926	9.914817	10.085183	35
30	9.803511	9.887406	9.916104	10.083895	30
35	9.804276	9.886885	9.917391	10.082609	25
40	9.805039	9.886362	9.918677	10.081323	20
45	9.805799	9.885837	9.919962	10.080038	15
50	9.806557	9.885311	9.921247	10.078753	10
55	9.807314	9.884783	9.922530	10.077470	5
60	9.808067	9.884254	9.923814	10.076186	0
	Co-fine.	Sine.	Co-tang.	Tangent.	M.

50.

The Table of Sines and Tangents.

40.

M.	Sine.	Co-sine.	Tangent.	Co-tangent.	
0	9.808067	9.884254	9.923814	10.076186	60
5	9.808819	9.883723	9.925096	10.074904	55
10	9.809569	9.883191	9.926378	10.073622	50
15	9.810316	9.882657	9.927659	10.072341	45
20	9.811061	9.882121	9.928940	10.071060	40
25	9.811804	9.881584	9.930220	10.069780	35
30	9.812544	9.881046	9.931499	10.068501	30
35	9.813283	9.880505	9.932778	10.067222	25
40	9.814019	9.879963	9.934056	10.065944	20
45	9.814753	9.879420	9.935333	10.064667	15
50	9.815485	9.878875	9.936611	10.063389	10
55	9.816215	9.878328	9.937887	10.062113	5
60	9.816943	9.877780	9.939163	10.060837	0
	Co-fine.	Sine.	Co-tang.	Tangent.	M.

49.

41.

M.	Sine.	Co-fine.	Tangent.	Co-tangent.	
0	9.816943	9.877780	9.939163	10.060837	60
5	9.817668	9.877230	9.940439	10.059561	55
10	9.818392	9.876678	9.941713	10.058287	50
15	9.819113	9.876125	9.942988	10.057012	45
20	9.819832	9.875571	9.944262	10.055738	40
25	9.820550	9.875014	9.945535	10.054465	35
30	9.821265	9.874456	9.946808	10.053192	30
35	9.821977	9.873896	9.948081	10.051919	25
40	9.822688	9.873335	9.949353	10.050647	20
45	9.823397	9.872772	9.950625	10.049375	15
50	9.824104	9.872208	9.951896	10.048104	10
55	9.824808	9.871641	9.953167	10.046833	5
60	9.825511	9.871073	9.954437	10.045563	0
	Co-fine.	Sine.	Co-tang.	Tangent.	M.

48.

The Table of Sines and Tangents.

42.

M.	Sine.	Co-sine.	Tangent.	Co-tangent.	
0	9.825511	9.871073	9.954437	10.045503	60
5	9.826211	9.870504	9.955708	10.044292	55
10	9.826910	9.869933	9.956977	10.043023	50
15	9.827616	9.869360	9.958247	10.041753	45
20	9.828301	9.868785	9.959516	10.040484	40
25	9.828993	9.868209	9.960784	10.039216	35
30	9.829683	9.867631	9.962052	10.037948	30
35	9.830372	9.867051	9.963320	10.036680	25
40	9.831058	9.866470	9.964588	10.035412	20
45	9.831742	9.865887	9.965855	10.034145	15
50	9.832425	9.865302	9.967123	10.032877	10
55	9.833105	9.864716	9.968389	10.031611	5
60	9.833783	9.864127	9.969656	10.030344	0
	Co-sine.	Sine.	Co-tang.	Tangent.	M.

47.

43.

M.	Sine.	Co-sine.	Tangent.	Co-tangent.	
0	9.833783	9.864127	9.969656	10.030344	60
5	9.834460	9.863538	9.970922	10.029078	55
10	9.835134	9.862946	9.972188	10.027812	50
15	9.835807	9.862353	9.973454	10.026546	45
20	9.836477	9.861758	9.974720	10.025280	40
25	9.837146	9.861161	9.975985	10.024015	35
30	9.837812	9.860562	9.977250	10.022750	30
35	9.838477	9.859962	9.978515	10.021485	25
40	9.839140	9.859360	9.979780	10.020220	20
45	9.839800	9.858756	9.981044	10.018956	15
50	9.840459	9.858151	9.982309	10.017691	10
55	9.841116	9.857543	9.983573	10.016427	5
60	9.841771	9.856934	9.984837	10.015163	0
	Co-sine.	Sine.	Co-tang.	Tangent.	M.

46.

The Table of Sines and Tangents.

44.

M.	Sine.	Co-sine.	Tangent.	Co-tangent.	
0	9.841771	9.856934	9.984837	10.015163	60
5	9.842424	9.856323	9.986101	10.013859	55
10	9.843076	9.855711	9.987365	10.012635	50
15	9.843725	9.855096	9.988629	10.011371	45
20	9.844372	9.854480	9.989893	10.010107	40
25	9.845018	9.853862	9.991156	10.008844	35
30	9.845662	9.853242	9.992420	10.007580	30
35	9.846304	9.852620	9.993683	10.006317	25
40	9.846944	9.851997	9.994947	10.005053	20
45	9.847582	9.851372	9.996210	10.003790	15
50	9.848218	9.850745	9.997473	10.002527	10
55	9.848852	9.850116	9.998737	10.001263	5
60	9.849485	9.849485	10.000000	10.000000	0
	Co-sine.	Sine.	Co-tang.	Tangent.	M.

45.

A

A TABLE
OF
Logarithm Numbers.

A Table of Logarithms.

N.	Logarith.	N.	Logarith	N.	Logarith.
1	0.000000	34	1.531479	67	1.826075
2	0.301030	35	1.544068	68	1.832509
3	0.477121	36	1.556303	69	1.838849
4	0.602060	37	1.568202	70	1.845098
5	0.698970	38	1.579783	71	1.851258
6	0.778151	39	1.591064	72	1.857332
7	0.845098	40	1.602060	73	1.863323
8	0.903090	41	1.612784	74	1.869232
9	0.954242	42	1.623249	75	1.875061
10	1.000000	43	1.633468	76	1.880813
11	1.041393	44	1.643452	77	1.886491
12	1.079181	45	1.653212	78	1.892094
13	1.113943	46	1.662758	79	1.897627
14	1.146128	47	1.672098	80	1.903090
15	1.176091	48	1.681241	81	1.908485
16	1.204120	49	1.690196	82	1.913814
17	1.230449	50	1.698970	83	1.919078
18	1.255272	51	1.707570	84	1.924279
19	1.278753	52	1.716003	85	1.929419
20	1.301230	53	1.724276	86	1.934498
21	1.322219	54	1.732394	87	1.939519
22	1.342422	55	1.740362	88	1.944482
23	1.361728	56	1.748188	89	1.949390
24	1.380211	57	1.755875	90	1.954242
25	1.397940	58	1.763428	91	1.959041
26	1.414973	59	1.770852	92	1.963788
27	1.431364	60	1.778151	93	1.968483
28	1.447158	61	1.785330	94	1.973128
29	1.462398	62	1.792391	95	1.977723
30	1.477121	63	1.799340	96	1.982271
31	1.491361	64	1.806180	97	1.986772
32	1.505150	65	1.812913	98	1.991226
33	1.518514	66	1.819544	99	1.995635
34	1.531479	67	1.826075	100	2.000000

A Table of Logarithms.

N.	Logarith.	N.	Logarith.	N.	Logarith.
101	2.004321	134	2.127105	167	2.222716
102	2.008600	135	2.130334	168	2.225309
103	2.012837	136	2.133539	169	2.227887
104	2.017033	137	2.136721	170	2.230449
105	2.021189	138	2.139879	171	2.232996
106	2.025306	139	2.143015	172	2.235528
107	2.029384	140	2.146128	173	2.238046
108	2.033424	141	2.149219	174	2.240549
109	2.037426	142	2.152288	175	2.243038
110	2.041393	143	2.155336	176	2.245513
111	2.045323	144	2.158362	177	2.247973
112	2.049218	145	2.161368	178	2.250420
113	2.053078	146	2.164353	179	2.252853
114	2.056905	147	2.167317	180	2.255273
115	2.060698	148	2.170262	181	2.257679
116	2.064458	149	2.173186	182	2.260071
117	2.068186	150	2.176091	183	2.262451
118	2.071882	151	2.178977	184	2.264818
119	2.075547	152	2.181844	185	2.267172
120	2.079181	153	2.184691	186	2.269513
121	2.082785	154	2.187521	187	2.271842
122	2.086359	155	2.190332	188	2.274158
123	2.089905	156	2.193125	189	2.276462
124	2.093422	157	2.195899	190	2.278754
125	2.096919	158	2.198657	191	2.281033
126	2.100371	159	2.201397	192	2.283301
127	2.103804	160	2.204110	193	2.285557
128	2.107209	161	2.206826	194	2.287802
129	2.110589	162	2.209515	195	2.290035
130	2.113943	163	2.212187	196	2.292256
131	2.117271	164	2.214844	197	2.294466
132	2.120574	165	2.217484	198	2.296665
133	2.123852	166	2.220108	199	2.298853
134	2.127105	167	2.222716	200	2.301029

200

A Table of Logarithms.

N.	Logarith.	N.	Logarith.	N.	Logarith.
201	2.303196	234	2.369216	267	2.426511
202	2.305351	235	2.371068	268	2.428135
203	2.307496	236	2.372912	269	2.429752
204	2.309630	237	2.374748	270	2.431364
205	2.311754	238	2.376577	271	2.432969
206	2.313867	239	2.378398	272	2.434569
207	2.315970	240	2.380211	273	2.436163
208	2.318063	241	2.382017	274	2.437751
209	2.320146	242	2.383815	275	2.439333
210	2.322219	243	2.385606	276	2.440909
211	2.324282	244	2.387389	277	2.442479
212	2.326336	245	2.389166	278	2.444045
213	2.328379	246	2.390935	279	2.445604
214	2.330414	247	2.392697	280	2.447158
215	2.332438	248	2.394452	281	2.448706
216	2.334454	249	2.396199	282	2.450249
217	2.336459	250	2.397940	283	2.451786
218	2.338456	251	2.399674	284	2.453318
219	2.340444	252	2.401401	285	2.454845
220	2.342422	253	2.403121	286	2.456366
221	2.344392	254	2.404834	287	2.457889
222	2.346353	255	2.406540	288	2.459392
223	2.348305	256	2.408239	289	2.460898
224	2.350248	257	2.409933	290	2.462398
225	2.352183	258	2.411619	291	2.463893
226	2.354108	259	2.413299	292	2.465383
227	2.356026	260	2.414973	293	2.466868
228	2.357935	261	2.416641	294	2.468347
229	2.359835	262	2.418301	295	2.469822
230	2.361728	263	2.419956	296	2.471292
231	2.363612	264	2.421604	297	2.472756
232	2.365488	265	2.423246	298	2.474216
233	2.367356	266	2.424882	299	2.475671
234	2.369216	267	2.426511	300	3.477121

A Table of Logarithms.

N.	Logarith.	N.	Logarith.	N.	Logarith.
301	2.478566	334	2.523746	367	2.564666
302	2.480007	335	2.525145	368	2.565848
303	2.481443	336	2.526339	369	2.567126
304	2.482874	337	2.527629	370	2.568202
305	2.484299	338	2.528916	371	2.569374
306	2.485721	339	2.530199	372	2.570543
307	2.487138	340	2.531479	373	2.571709
308	2.488551	341	2.532754	374	2.572872
309	2.489958	342	2.534026	375	2.574031
310	2.491362	343	2.535294	376	2.575188
311	2.492760	344	2.536558	377	2.576341
312	2.494155	345	2.537819	378	2.577492
313	2.495544	346	2.539076	379	2.578639
314	2.496929	347	2.540329	380	2.579784
315	2.498311	348	2.541579	381	2.580925
316	2.499687	349	2.542825	382	2.582063
317	2.501059	350	2.544068	383	2.583199
318	2.502427	351	2.545307	384	2.584331
319	2.503791	352	2.546543	385	2.585461
320	2.505149	353	2.547775	386	2.586587
321	2.506505	354	2.549003	387	2.587711
322	2.507856	355	2.550228	388	2.588832
323	2.509203	356	2.551449	389	2.589949
324	2.510545	357	2.552668	390	2.591065
325	2.511883	358	2.553883	391	2.592177
326	2.513218	359	2.555094	392	2.593286
327	2.514548	360	2.556303	393	2.594393
328	2.515874	361	2.557507	394	2.595496
329	2.517196	362	2.558709	395	2.596597
330	2.518514	363	2.559907	396	2.597695
331	2.519818	364	2.561101	397	2.598790
332	2.521138	365	2.562293	398	2.599883
333	2.522444	366	2.563481	399	2.600973
334	2.523746	367	2.564666	400	2.602059

A Table of Logarithms.

N.	Logarith.	N.	Logarith.	N.	Logarith.
401	2.603144	434	2.637489	467	2.669317
402	2.604226	435	2.638489	468	2.670246
403	2.605305	436	2.639486	469	2.671173
404	2.606381	437	2.640481	470	2.672098
405	2.607455	438	2.641475	471	2.673021
406	2.608526	439	2.642469	472	2.673942
407	2.609594	440	2.643453	473	2.674861
408	2.610660	441	2.644439	474	2.675778
409	2.611723	442	2.645422	475	2.676694
410	2.612784	443	2.646404	476	2.677607
411	2.613842	444	2.647383	477	2.678518
412	2.614897	445	2.648360	478	2.679428
413	2.615950	446	2.649335	479	2.680336
414	2.617000	447	2.650308	480	2.681241
415	2.618048	448	2.651278	481	2.682145
416	2.619093	449	2.652246	482	2.683047
417	2.620136	450	2.653213	483	2.683947
418	2.621176	451	2.654177	484	2.684845
419	2.622214	452	2.655138	485	2.685742
420	2.623249	453	2.656098	486	2.686636
421	2.624282	454	2.657056	487	2.687529
422	2.625312	455	2.658011	488	2.688419
423	2.626340	456	2.658965	489	2.689309
424	2.627366	457	2.659916	490	2.690196
425	2.628389	458	2.660865	491	2.691081
426	2.629409	459	2.661813	492	2.691965
427	2.630428	460	2.662758	493	2.692847
428	2.631444	461	2.663701	494	2.693727
429	2.632457	462	2.664642	495	2.694605
430	2.633468	463	2.665581	496	2.695482
431	2.634477	464	2.666518	497	2.696356
432	2.635484	465	2.667453	498	2.697229
433	2.636488	466	2.668386	499	2.698101
434	2.637489	467	2.669317	500	2.698970

A Table of Logarithms.

N.	Logarith.	N.	Logarith.	N.	Logarith.
501	2.699838	534	2.727541	567	2.753583
502	2.700704	535	2.728354	568	2.754348
503	2.701568	536	2.729165	569	2.755112
504	2.702430	537	2.729974	570	2.755875
505	2.703291	538	2.730782	571	2.756636
506	2.704151	539	2.731589	572	2.757396
507	2.705008	540	2.732394	573	2.758155
508	2.705863	541	2.733197	574	2.758912
509	2.706718	542	2.733999	575	2.759668
510	2.707570	543	2.734799	576	2.760422
511	2.708421	544	2.735599	577	2.761176
512	2.709269	545	2.736397	578	2.761928
513	2.710117	546	2.737192	579	2.762679
514	2.710963	547	2.737987	580	2.763428
515	2.711807	548	2.738781	581	2.764176
516	2.712649	549	2.739572	582	2.764923
517	2.713491	550	2.740363	583	2.765669
518	2.714329	551	2.741152	584	2.766413
519	2.715167	552	2.741939	585	2.767156
520	2.716003	553	2.742725	586	2.767898
521	2.716838	554	2.743509	587	2.768638
522	2.717671	555	2.744293	588	2.769377
523	2.718502	556	2.745075	589	2.770115
524	2.719331	557	2.745855	590	2.770852
525	2.720159	558	2.746634	591	2.771587
526	2.720986	559	2.747412	592	2.772322
527	2.721811	560	2.748188	593	2.773055
528	2.722634	561	2.748963	594	2.773786
529	2.723456	562	2.749736	595	2.774517
530	2.724276	563	2.750508	596	2.775246
531	2.725095	564	2.751279	597	2.775974
532	2.725912	565	2.752048	598	2.776701
533	2.726727	566	2.752816	599	2.777427
534	2.727541	567	2.753583	600	2.778151

A Table of Logarithms.

N.	Logarith.	N.	Logarith.	N.	Logarith.
601	2.778874	634	2.802089	667	2.824126
602	2.779596	635	2.802774	668	2.824776
603	2.780317	636	2.803457	669	2.825426
604	2.781037	637	2.804139	670	2.826075
605	2.781755	638	2.804821	671	2.826723
606	2.782473	639	2.805501	672	2.827369
607	2.783189	640	2.806179	673	2.828015
608	2.783904	641	2.806858	674	2.828659
609	2.784617	642	2.807535	675	2.829304
610	2.785329	643	2.808211	676	2.829947
611	2.786041	644	2.808886	677	2.830589
612	2.786751	645	2.809559	678	2.831229
613	2.787460	646	2.810233	679	2.831869
614	2.788164	647	2.810904	680	2.832509
615	2.788875	648	2.811575	681	2.833147
616	2.789581	649	2.812245	682	2.833784
617	2.790285	650	2.812913	683	2.834421
618	2.790988	651	2.813581	684	2.835056
619	2.791691	652	2.814248	685	2.835691
620	2.792392	653	2.814913	686	2.836324
621	2.793092	654	2.815578	687	2.836957
622	2.793791	655	2.816241	688	2.837588
623	2.794488	656	2.816904	689	2.838219
624	2.795185	657	2.817565	690	2.838849
625	2.795880	658	2.818226	691	2.839478
626	2.796574	659	2.818885	692	2.840106
627	2.797268	660	2.819543	693	2.840733
628	2.797959	661	2.820201	694	2.841359
629	2.798651	662	2.820858	695	2.841985
630	2.799341	663	2.821514	696	2.842609
631	2.800029	664	2.822168	697	2.843233
632	2.800717	665	2.822822	698	2.843855
633	2.801404	666	2.823474	699	2.844477
634	2.802089	667	2.824126	700	2.845098

A Table of Logarithms.

N.	Logarith.	N.	Logarith.	N.	Logarith.
701	2.845718	734	2.865696	767	2.884795
702	2.846337	735	2.866287	768	2.885361
703	2.846955	736	2.866878	769	2.885926
704	2.847573	737	2.867467	770	2.886491
705	2.848189	738	2.868056	771	2.887054
706	2.848805	739	2.868643	772	2.887617
707	2.849419	740	2.869232	773	2.888179
708	2.850033	741	2.869818	774	2.888741
709	2.850646	742	2.870404	775	2.889302
710	2.851258	743	2.870989	776	2.889862
711	2.851869	744	2.871573	777	2.890421
712	2.852479	745	2.872156	778	2.890979
713	2.853089	746	2.872739	779	2.891537
714	2.853698	747	2.873321	780	2.892095
715	2.854306	748	2.873902	781	2.892651
716	2.854913	749	2.874482	782	2.893207
717	2.855519	750	2.875061	783	2.893762
718	2.856124	751	2.875639	784	2.894316
719	2.856729	752	2.876218	785	2.894869
720	2.857332	753	2.876795	786	2.895423
721	2.857935	754	2.877371	787	2.895975
722	2.858537	755	2.877947	788	2.896526
723	2.859138	756	2.878522	789	2.897077
724	2.859739	757	2.879096	790	2.897627
725	2.860338	758	2.879669	791	2.898176
726	2.860937	759	2.880242	792	2.898725
727	2.861534	760	2.880814	793	2.899273
728	2.862131	761	2.881385	794	2.899821
729	2.862728	762	2.881955	795	2.900367
730	2.863323	763	2.882525	796	2.900913
731	2.863917	764	2.883093	797	2.901458
732	2.864511	765	2.883661	798	2.902003
733	2.865104	766	2.884229	799	2.902547
734	2.865696	767	2.884795	800	2.903089

D 3

A Table of Logarithms.

N.	Logarith.	N.	Logarith.	N.	Logarith.
801	2.903633	834	2.921166	867	2.938019
802	2.904174	835	2.921686	868	2.938519
803	2.904716	836	2.922206	869	2.939019
804	2.905256	837	2.922725	870	2.939519
805	2.905796	838	2.923244	871	2.940018
806	2.906335	839	2.923762	872	2.940516
807	2.906874	840	2.924279	873	2.941014
808	2.907411	841	2.924796	874	2.941511
809	2.907949	842	2.925312	875	2.942008
810	2.908485	843	2.925828	876	2.942504
811	2.909021	844	2.926342	877	2.942999
812	2.909556	845	2.926857	878	2.943495
813	2.910091	846	2.927370	879	2.943989
814	2.910624	847	2.927883	880	2.944483
815	2.911158	848	2.928396	881	2.944976
816	2.911690	849	2.928908	882	2.945468
817	2.912222	850	2.929419	883	2.945961
818	2.912753	851	2.929929	884	2.946452
819	2.913284	852	2.930439	885	2.946943
820	2.913814	853	2.930949	886	2.947434
821	2.914343	854	2.931458	887	2.947924
822	2.914872	855	2.931966	888	2.948413
823	2.915399	856	2.932474	889	2.948902
824	2.915927	857	2.932981	890	2.949390
825	2.916454	858	2.933487	891	2.949878
826	2.916980	859	2.933993	892	2.950365
827	2.917506	860	2.934498	893	2.950851
828	2.918030	861	2.935003	894	2.951338
829	2.918555	862	2.935507	895	2.951823
830	2.919078	863	2.936011	896	2.952308
831	2.919601	864	2.936514	897	2.952792
832	2.920123	865	2.937016	898	2.953276
833	2.920645	866	2.937518	899	2.953759
834	2.921166	867	2.938019	900	2.954243

A Table of Logarithms.

N.	Logarith.	N.	Logarith.	N.	Logarith.
901	2.954725	934	2.970347	967	2.985426
902	2.955207	935	2.970812	968	2.985875
903	2.955688	936	2.971276	969	2.986324
904	2.956168	937	2.971739	970	2.986772
905	2.956649	938	2.972203	971	2.987219
906	2.957128	939	2.972666	972	2.987666
907	2.957607	940	2.973128	973	2.988113
908	2.958086	941	2.973589	974	2.988559
909	2.958564	942	2.974050	975	2.989005
910	2.959041	943	2.974512	976	2.989449
911	2.959518	944	2.974972	977	2.989895
912	2.959995	945	2.975432	978	2.990339
913	2.960471	946	2.975891	979	2.990783
914	2.960946	947	2.976349	980	2.991226
915	2.961421	948	2.976808	981	2.991669
916	2.961895	949	2.977266	982	2.992111
917	2.962369	950	2.977724	983	2.992554
918	2.962843	951	2.978181	984	2.992995
919	2.963315	952	2.978637	985	2.993436
920	2.963788	953	2.979093	986	2.993877
921	2.964259	954	2.979548	987	2.994317
922	2.964731	955	2.980003	988	2.994756
923	2.965202	956	2.980458	989	2.995196
924	2.965672	957	2.980912	990	2.995635
925	2.966142	958	2.981366	991	2.996074
926	2.966611	959	2.981819	992	2.996512
927	2.967079	960	2.982271	993	2.996949
928	2.967548	961	2.982723	994	2.997386
929	2.968016	962	2.983175	995	2.997823
930	2.968483	963	2.983626	996	2.998259
931	2.968949	964	2.984077	997	2.998695
932	2.969416	965	2.984527	998	2.999130
933	2.969882	966	2.984977	999	2.999565
934	2.970347	967	2.985426	1000	3.000000

The Use of these TABLES hath been already at large shewed in the First and Twelfth Chapters; therefore I shall say no more of them here.

A N

AN APPENDIX,

Shewing farther

How to Survey by the CHAIN only: With an useful TABLE to that Purpose.

HAVING, in the fixth Chapter of the foregoing Treatife, taught a ready and eafy Way for taking the quantity of an Angle in the Field by the Chain only; and underftanding it to have met with good Acceptance among Surveyors and others: I thought it proper to fay fomething more on that Subject, the prefent Opportunity of a new Edition of the Book inviting me thereto. And that this way of working may be practifed as quick and true as any in the World, with all the coftly Inftruments that ever were invented, there are two feeming Difficulties to be removed. The *firſt* is, when the Angle grows very obtufe, that is to fay, containing 170 Degrees or more, then the Subtendent or Chord-Line will hardly be diftinguifhable between five or fix Degrees, there being but $\frac{1}{7}$ Part of a Link difference between 170 Deg. and 171 Degrees, and not above $\frac{1}{75}$ Part, between 178, and 179 Degrees. To remedy which, you need not take the Quantity of that Angle at all, efpecially if it be an inward Angle, but meafure directly from B to C, and when you come right againft A, take an Off-fet (which you may do with a Rod or Line alone, as true as with a

Crofs

(2)

Crofs or other Inftrument) which Off-fet, put down in your Field-Book, will do the Bufinefs when you come to protract, as well as if you had taken the Angle in the Field : But if that does not pleafe you, or any other Reafon neceffitate you to take the Angle *A*, there place a ftrong Stick in the very Angle *A*, and putting the Ring of the Chain over it, ftretch it out at full length, both in the Line *A B* and *A C*, and where the end of the Chain falls, there place Sticks alfo, as at *D* and *G*. Remove your Chain from *A*, and put the Ring over the Stick at *D*, and ftretch it out at adventure as towards *E*. Now you fhould have another Chain or a fmall Line (which you may carry in your Pocket) exactly of the length of a Chain, with a Loop at each End, which put over the Stick at *A*, and taking the other Loop of the Line in one Hand, and the loofe End of the Chain in the other Hand, go backward, till both being ftretched ftrait, meet at *E*, then have you found *D A E*, an Equilateral Triangle confifting of 60 Degrees; to which add another Equilateral Triangle by loofing the Chain at *D*, and putting it over the Stick at *E*, letting the Line remain as it was faftened at *A*, and taking the loofe Ends again of the Chain and Line in your Hands, go backwards as before, until both being ftretched ftrait, meet in *F*. So have you found two Equilateral Triangles, or 120 Degrees. Laftly, with your Chain, meafure the neareft Diftance *FG*, which fuppofe to be 84 Links and a half, which Sum look for in the following Table, and right againft it you will find 50 Degrees, which added to 120, make

(3)

make 170 Degrees, the Quantity of the Angle fought, or if you have not a Mind to use the Table, you may note it down in your Field-Book, thus, △ △ 84 ½, signifying that Angle consists of two Equilaterals and 84 ½ Links for its Subtendent, and you may plot it by doing with your Compasses upon the Paper, what you did in the Field with the Chain. But now perhaps you may be ready to say, You pretend to teach how to take the Quantity of an obtuse Angle with the Chain only, and here is a Line required, or two Chains at least. Well then, you shall presently see how to do it with one Chain only. Let *E A I* in the following Figure, be the Angle of 170 Degrees, measure from *A* towards *B* and *C*, half a Chain on each Side, as to *D* and *E*, where stick down Sticks, and one at *A*, then put the Ring at one end of the Chain over the Stick at *A*, and the other end over the Stick at *D*, and taking the Chain in the middle by the Ring that is commonly at the End of 50 Links, go backwards till both Parts are strait, and there stick down a Stick, as at *F*. Then loose the Ring from *D*, and put it over the Stick at *F*, and taking the very middle of the Chain, make both Parts strait, which they will be at *C*, where stick down a Stick, from which measure to *E*, noting it down in your Field-Book △ △ 42 ½, and when you plot it, remember to make your Equilaterals but of 50 Links the Sides of them: I say, when you plot it, for you may not in this Case have recourse to the following TABLE, that being made to the

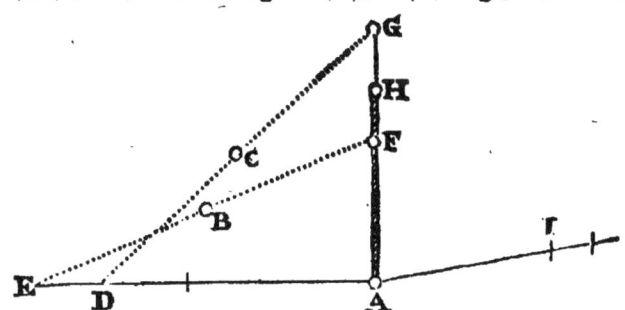

Radius of 100 Links, unless you double the number of Links found between *C* and *E*; or, which is better, when

you

you have finish'd your two Equilaterals, one end of the Chain hanging at *A*, stretch the other at full length over the Stick at *G*, which will fall at *H*, then measuring the nearest Distance between *H* and *C*, you will find it to be 84 1/2 Links, against which in the Table stand 50 Degrees, which added to your two Equilaterals, make 170 Degrees for the Angle *A*. [*See this Figure.*]

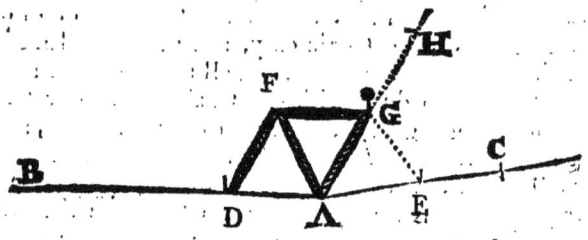

But now if you had rather measure this Angle, by first taking out a right Angle from it: Thus you may do to find the Perpendicular for the right Angle: [*See the Figure on the other Side.*] Put one Ring of your Chain over the Stick at the Angle *A*, and stretching out the Chain, let the other end fall any where at adventure, as at *B* or *C*, where stick a Stick through the Ring, and loosing that End at *A*, take it in your Hand, and stretching it strait, see in what Part it will just touch the Hedge *A E*, which will be at *D*, if the other End be at *C*; or at *E*, if the other End be at *B*; and there make a Mark, which done, keeping the End of the Chain in your Hand go backward from *B* or *C*, towards *G* or *F*, till your Chain is strait, then moving your self side ways to and fro, till you perceive your Chain to lie in a strait Line with *B E*, or *C D*, at the End of it place a Stick, as at *F* or *G*, from whence to *A* will be a Line perpendicular to *A E*, wherefore from *A* set off one Chain in that Line which will fall at *H*, and one Chain upon the Line *A I*, which falls at *I*, and measuring the Distance *H I*, you will find it 128 Links 7/12 Parts of a Link, or 80 Deg. which added to the right Angle makes 170 Deg. which was the Angle required.

Otherwise you may take a Right-Angle, by fixing one End of the Chain in the Angle it self, and the other End at 40 Links distance in the Hedge, then take 50 Links in
one

one hand, and 30 in the other, and ſtretch both Parts ſtrait; their Meeting will conſtitute a Right Angle according to the well known Axiom, that 3, 4, and 5 make a Right-Angled Triangle.

Many other Ways might be ſhewn, to take a right Angle in the Field by the Chain only, as alſo otherwiſe to meaſure the Quantity of an obtuſe Angle; but I omit them, leaving it to your own Practice and Ingenuity: Only one Way more and the very beſt, to take the Quantity of this obtuſe Angle, which take as follows.

In the following Figure let *A* be the Angle required to be taken in the Field; by the Chain firſt from *A*, ſet off two Chains, one to *B*, the other to *C*, then fixing one End of the Chain in *B*, ſtretch the other direct in a ſtrait Line towards *C*, making a Mark where the End falls, which will be at 7; meaſure the Diſtance from 7 to *A*, which ſuppoſe to be 8 Links $\frac{7}{10}$ Parts of a Link, look in the following TABLE, and right againſt it you will find

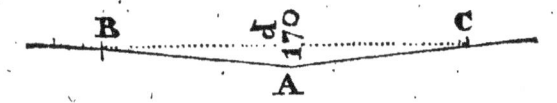

$$B \quad\quad\quad\quad \cdots\cdots\cdots 7 \cdots\cdots\cdots \quad\quad\quad\quad C$$
$$A$$

5 Degrees, which doubled (the Angles *A C* 7, and *A B* 7, being equal, becauſe the Sides *A B* and *A C* are equal, and it) makes 10 Degrees, which ſubtracted from 180, leaves 170 for the deſired Angle at *A*.

But now if this had been an outward Angle, as we will ſuppoſe the following; you have no more to do, but to continue one Line, as for Example, the Line *D A* to *C*, one Chain, and alſo to ſet off one Chain upon the other Line from *A* to *B*, then meaſure the Diſtance *B C*, which

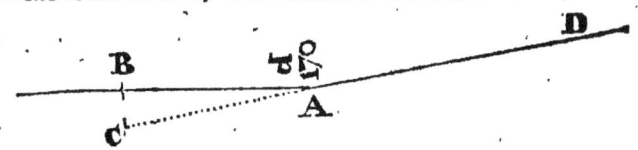

$$\quad\quad\quad\quad\quad\quad\quad\quad\quad\quad\quad\quad D$$
$$B \quad\quad\quad 7 \quad\quad\quad\quad$$
$$C \cdots\cdots\cdots A$$

ſay to be 17 Links $\frac{4}{10}$ Parts of a Link, which anſwers in the Table to 10 Degrees, which is the Complement of the Angle *A* to 180 Degrees; wherefore take 10 from 180, remains 170 for the ſought outward Angle.

By

By this time, I hope, the Difficulty of measuring obtuse Angles is well removed, and the Matter made plain and easy: As for acute Angles, and such obtuse ones as are but a little bigger than 90 Degrees, you have the Way to measure them already in the 6th Chapter of the foregoing Treatise; with sundry Ways to measure a Field with the Chain only, to which I refer you.

It remains now to speak of the second seeming Difficulty, which lies in the Trouble of plotting after this way: To remove which, you may have a Protractor made with Links on it instead of Degrees, or both, if you please, which the Instrument-maker may soon do, by the help of this Table. Or you may very well use your ordinary Protractors; for having a Copy of this Table in the Field with you, you may at once note down the Degrees of every Angle, without mentioning the Subtendents at all, or if you do only note down the Subtendents in your Fieldbook, when you come home, you may at once take all the Angles in Degrees answerable to them, and so plot with an ordinary Protractor, as at other times. I have made the Table but to 140 Degrees; for as I told you before, when an Angle exceeds that, your best way of measuring it, is as has been just now taught.

What has been already said I presume to be sufficient to explain the following Table, and the Use thereof, therefore shall not trouble you with Repetitions; only desire you to remember, that the Table is made for the Radius of one Chain, or 100 Links: And the Subtendents, or Chord Lines are in Links, and decimal Parts of a Link: So that when you would use this Table, you must set off but one Chain from the Angle (you desire to know the Quantity of) on either Hedge, and measuring the nearest Distance between the two Ends of the Chains a-cross from Hedge to Hedge, look for the Number of Links in the Table that nearest Distance contains, and right against it you will find the Quantity of the Angle as true, if not truer, than if it had been taken by the best *Semicircle, Circumferentor,* or *Theodolite.*

EXAMPLE.

(7)
EXAMPLE.

In Folio 113 of the foregoing Book, I would know the Quantity in Degrees of the Angle *eE*, whose Subtendent is there (accounting one Chain Radius) said to be 80 Links: Accordingly I look for 80 Links in the Table, and the nearest Number to it is 79 Links $\frac{2}{14}$ Parts of a Link, and right against it stand 47 Degrees: Wherefore I say that Angle consists of 47 Degrees, and a little more; and tho' it be needless, yet if you desire to know how much that odd $\frac{1}{14}$ is (which is wanting to make up 80 Links) you may see by the Table, that in an Angle of this Bigness one Link and half raises a Degree; so that $\frac{1}{14}$ Parts of a Link is just 12 Minutes. The exact Angle therefore is 47° 12'.

What has been said concerning Measuring a Field, or taking an Angle by the Chain only, either in the APPENDIX, or Sixth Chapter, may as well be applied to a Pole or Rod cut out of the Hedge and divided into 100 equal Parts; and indeed you may altogether as well, and much quicker, do it with a Rod than the Chain, every Division of the Rod answering to a Link of the Chain: But then you must take care your Rod be strait; and the Table serves as well for a Rod so divided as the Chain, only in casting up there is a Difference (which your own Reason and the foregoing Treatise will sufficiently explain to you) unless in Measuring the Length of the Lines, you call every 4 Poles 1 Chain, and every 4 Divisions of the Pole 1 Link; then you may cast it as if it had been measured by the Chain: But there is no need for that, that I know of. You may have, I suppose, in *Crooked-Lane*, a Rod made to shoot one Part into another like a Fishing-rod, to be used as a Cane, in the Head whereof may be a small Compass; which alone is Instrument enough to survey any Piece of this Earth, be it Mannor or larger: And if so, what need is there of a Horse-load of Brass Circles, and Semicircles, heavy Ball-Sockets, wooden Tables and Frames, and 3 legged Staffs, *cum multis aliis*, unless to amuse the ignorant Countryman, to make him more freely pay the Surveyor?

The

The TABLE of Chords, or Subtendents to the Radius of one Chain of Gunter's, br 100 Links.

Degrees.	Links.	Tenths of a Link.	Degrees.	Links.	Tenths.	Degrees.	Links.	Tenths.	Degrees.	Links.	Tenths.
1	1	.7	36	61	.8	71	116	.1	106	159	.7
2	3	.5	37	63	.4	72	117	.5	107	160	.8
3	5	.2	38	65	.1	73	119	.0	108	161	.8
4	7	.0	39	66	.8	74	120	.4	109	162	.8
5	8	.7	40	68	.4	75	121	.8	110	163	.8
6	10	.5	41	70	.0	76	123	.1	111	164	.8
7	12	.2	42	71	.7	77	124	.5	112	165	.8
8	14	.0	43	73	.3	78	125	.9	113	166	.8
9	15	.7	44	74	.9	79	127	.2	114	167	.7
10	17	.4	45	76	.5	80	128	.5	115	168	.7
11	19	.2	46	78	.1	81	129	.9	116	169	.6
12	20	.9	47	79	.7	82	131	.2	117	170	.5
13	22	.6	48	81	.3	83	132	.5	118	171	.4
14	24	.4	49	82	.9	84	133	.8	119	172	.3
15	26	.1	50	84	.5	85	135	.1	120	173	.2
16	27	.8	51	86	.1	86	136	.4	121	174	.1
17	29	.6	52	87	.7	87	137	.7	122	174	.9
18	31	.3	53	89	.2	88	139	.0	123	175	.7
19	33	.0	54	90	.8	89	140	.2	124	176	.6
20	34	.7	55	92	.3	90	141	.4	125	177	.4
21	36	.4	56	93	.9	91	142	.6	126	178	.2
22	38	.2	57	95	.4	92	143	.8	127	179	.0
23	39	.9	58	97	.0	93	145	.0	128	179	.8
24	41	.6	59	98	.5	94	146	.2	129	180	.5
25	43	.3	60	100	.0	95	147	.4	130	181	.3
26	44	.9	61	101	.5	96	148	.6	131	182	.0
27	46	.7	62	103	.0	97	149	.8	132	182	.7
28	48	.4	63	104	.5	98	151	.0	133	183	.4
29	50	.1	64	106	.0	99	152	.1	134	184	.0
30	51	.8	65	107	.4	100	153	.2	135	184	.7
31	53	.4	66	108	.9	101	154	.3	136	185	.4
32	55	.1	67	110	.4	102	155	.4	137	186	.1
33	56	.8	68	111	.8	103	156	.5	138	186	.7
34	58	.5	69	113	.3	104	157	.6	139	187	.3
35	60	.1	70	114	.7	105	158	.7	140	187	.9

FINIS.